EYES ON THE WILDERNESS

EYES
ON THE
WILDERNESS

HELMUT HIRNSCHALL

 HANCOCK HOUSE

ISBN 0-919654-39-8

Copyright © 1975   Helmut Hirnschall

Canadian Shared Cataloguing in Publication Data

Hirnschall, Helmut
    Eyes on the wilderness

    1. Animals, Habits and behavior of.
2. Birds - Behavior. · I. Title.

QL751.H57   591.5
ISBN 0-919654-39-8

Designed by Bob Wilcox
Type set in Times New Roman
Cover design by Don Matheson

Printed in U.S.A.

This book was designed and first produced in Canada by
Hancock House Publishers Limited, 3215 Island View Road,
Saanichton, British Columbia, Canada.

HANCOCK HOUSE PUBLISHERS, LTD.
3215 Island View Road
Saanichton, British Columbia Canada

CONTENTS

ACKNOWLEDGMENTS

Since I settled in Canada in 1960 I have learned to enjoy and love a life I have never known before. This love has been tested over the years with trials and tribulations, all of which were rendered insignificant by my wilderness experiences and by the many people who did their best to make me feel at home in a new country and a new language. In their own ways these people have, in a sense, become my collaborators for this book. Some have guided my first steps in exploring the wilderness, others have offered hospitality, still others lasting friendship. To some I am indebted for an insight, a view, an explanation, or perhaps just a thought. G. Beerkens taught me the art of kayaking and wilderness comradeship. G. Dokis introduced me to Indian hospitality and told me the first Indian legends. A game guide from Golden gave me a compassionate view on wolves, and a friendly old man whose last name I have never known has been a lasting inspiration in my desire to stay close to Nature. I think of these people, and many others, with lasting gratitude.

Many books have been written about animals in a factual, scientific manner, and therefore I saw no need to write another one. Lacking academic training, I felt less burdened with facts that often obscure the naive curiosity to gather basic knowledge, and wrote this book as a personal testimony to my love for the Canadian wilderness. I am confessing to this love with a childish impudence, and dedicate this book to all those who suffer from the same affliction.

My special thanks go to John Moelaert for his editorial expertise and to Joanna and Mark for a companionship that is most conducive to my creative urges. Lastly, I feel compelled to express my thanks to the chipmunk, the beaver, the bear, and all other animals that roam the wild. Then, of course, there are the clouds, the sky, the trees...

Helmut Hirnschall

FOREWORD

For untold generations my people have lived in this country—each man in harmony with his surroundings. To us Nature was a church—it was revered, loved, respected, and worshipped. We never took from it without asking the Great Spirit's permission in prayer and always thanked him for giving us of the goodness which he offered in Nature so freely.

Now all this is changed.

As the ever increasing demands by mankind diminish the world's wilderness areas, modern man now has less of a chance to experience Nature firsthand. Unless we go back to the old ways of loving and caring for Nature it will not be preserved and some day the dreadful realization will come to us that without it man cannot exist.

We are still fortunate to find unspoiled wilderness areas in this country that tell us how the world was thousands of years ago. This same land thrives and throbs with the progressiveness of people that seek to achieve the status of the most advanced society.

Let us not forget to acknowledge Nature's place in such a society. Let us not forget that each of us has a common bond with the cedar seed that grows into a huge tree, or with the bear cub that needs a mother's love to become a strong and able being. Let us not forget that we need clean air for our lungs and pure water to quench our thirst. Let us not forget that we are not the last generation to live on earth and that humans are not her only residents. Let us not forget that we are but visitors on this earth and therefore, like the conscientious businessman, who puts his affairs in order before passing them on to his successor, we too must endeavor to leave this world in an orderly state so that our children will remember us with gratitude and warmth and marvel at the beauty of an unchanging Nature.

If we are unable to experience true wilderness adventures ourselves we can at least get some share of it by reading those books that have been written by someone who loves and cares about Nature.

It is my hope that you will find some of this love in the pages of this book. It is my hope that reading it will stimulate you to discover for yourself the great, quiet beauty of our Canadian wilderness heritage and help preserve it. It is my hope that this book will show you the value of living in peace with Nature.

Chief Dan George
June, 1975
Burrard Reserve, Vancouver, B.C.

CHILDHOOD "WILDERNESS"

North American wildlife has had a fascination for me ever since I came to the continent in 1960, a 27-year-old immigrant from Austria. At least a part of my attraction to wilderness animals formed after encountering for the first time that most recognizable of North American mammals, the beaver. So completely different from any so-called 'wild animal' I had ever known in my homeland, my curiosity about the beaver, its lifestyle and habitat, led me eventually to study all other North American animals with as much eagerness as I had displayed in my youth.

At the age of eight a particular experience shaped my lifetime attitude towards animals.

My mother, who had lost her husband, did her best to raise two boys and a girl with the help of her mother and whatever a vegetable garden and four hens would yield. One of the hens had reached the end of her egg-laying days and now was facing her alternate fate: chicken stew.

My grandmother volunteered to play the unpleasant role of executioner and told me to catch the victim. I quickly chased the hen into the tool shed, where I cornered and caught her. Having done this I told my grandmother that I was not only capable of catching a hen but also of killing one. This, of course, was a vain attempt to grow up a little faster. But the wood chopping block was too high for me, and the axe was too heavy. Grandmother was certain that if I had my way the hen would only be injured and would

11

suffer needlessly. Instead I was to hold the hen's head so it would not drop to the ground.

Grandmother placed the hen belly-up on the block, restraining her with one hand, while the other gripped the axe. I stepped closer, and the hen twisted her neck and looked at me. As I placed my hands around her head I felt a weakness creep through my body.

Sometimes lack of experinece robs us of the opportunity to interfere in an event of which we do not approve. Before I could say: "Maybe...," the axe came down and I felt the weight of the hen's head in my cupped hands with a suddenness that sent me into shock. The body escaped my grandmother's hold and flew against the boards and shelves. The head in my hands grew heavier and heavier, while the eyes kept looking at me.

Eventually the flutter subsided, the body tumbled to the ground and staggered about spattering blood. I dropped the head and then ran to the far corner of the garden where I became violently sick.

At dinner time I refused to eat, but no undue pressure was put on me to change my mind. Since meat was a rarity on my mother's menu, it was not often thereafter that my memory dictated a refusal of chicken dishes. In time, the growing reasoning power of advancing years helped me to re-appreciate meat dishes, but to this day I abhor the killing of an animal.

I was lucky enough to have been brought up in a small town where most of my school friends lived on small farms. This provided me with many opportunities to meet farm animals and on two other occasions I unwittingly contributed to the death of goslings and rabbits.

Once a friend showed me four newly hatched goslings in his barn, that looked like they had been dipped into egg yolk. Their ruffled down gave them a fluffy loveliness that intrigued us so much that we carried them into the yard to watch them in the sunlight. Since they failed to amuse us by waddling around, we thought they might like to swim. With protective care we ushered them into the trough that was always running over with cool water from the well, and taught them all we knew. There is not much one can teach goslings that are only a few hours old. They tire easily—a conclusion we reached when one of the others closed its eyes. We carried them back to their nest and let them have their well deserved rest carefully tucking them in. Our care and compassion was completely misconstrued as mischief by my friend's mother when she later found the gosling had died of exposure. Within minutes of the tragic discovery the resolute woman acted as coroner, judge, jury, and executioner. With a sore behind, some advice and a new experience to look back on, I was sent home.

At another time, when it was my duty to feed young rabbits which had been added to our livestock for economic reasons, I came upon a pile of leaves under an apple tree. I decided that they were just as

good a rabbit food as the grass I was supposed to fetch, and thus stuffed their cage full of leaves. Some hours later three of the six rabbits had choked and were lying in their little beds of leaves with stems sticking out of their mouths. My mother, misunderstanding my inventiveness, enlightened me with a large wooden spoon on the same part of my anatomy that throughout the years became her favorite target for corporal punishment.

Aside from chickens, rabbits, a mongrel dog, and some stray cats I could never catch, there were no other animals at home. Whenever I was invited to my friend's farm I was fascinated by geese, ducks, turkeys, sheep, goats, pigs, calves, sows, steers, bulls, and horses. Though I was very eager to get involved with animals, I was also quite scared of them—especially the larger ones. Whenever I was to help my friend with the handling of animals, I insisted on a demonstration. Once he asked me to ride their plow horse into the barn. I was afraid of the big animal. My friend climbed onto its broad back to show me how harmless the horse was. He beat the hefty creature with a rope, but it would not move. This convinced me that large animals only look fierce and therefore encouraged me to act more boldly.

Cows seemed to be too docile to be dangerous. After all, they don't even interrupt their meal when approached by someone. Once I saw my friend stroke a front leg of a cow. I wanted to show him how much of a farm boy I had become, so I walked over to the next cow, knelt beside her and patted her belly. I had seen him do it many times. "Appearances can be deceiving" was the immediate painful lesson she taught me with one of her hoofs. She slugged me across the face and then, while I lay on the ground sobbing, stared at me with eyes big enough to reflect the whole barn, the fields and the trees. She never stopped chewing.

These experiences and the lack of wildlife in Austria gave me the impression that domestic animals were wild beasts. This attitude was reinforced by the literature we had to read in school. The books were full of stories describing how ancestors fought with wolves and bears.

When Caesar and his troops reached the Danube they saw on the other side "impenetrable forests, as black as the night." They left those forests to the bears, the wolves, the deer, the wildcats, the eagles, the wild boars, and all other beasts, and established their forts on the south side of the river.

Two-thousand years later one of these forts called Vindobona had grown into a city of worldwide fame: Vienna. The capital of Austria is now the home of two million people, most of whom have never experienced the darkness that lurks under wild grown trees, the coolness of a giant cedar in the midsummer sun, the

14

snarl of a lynx, or the pureness of water from an untamed river. Over the centuries the wilderness south and north of the Danube has been reduced to pitiful little clusters of trees called forests. The land has been changed into a quilted patchwork of small farms. In many places Austria's Rivers have been robbed of their former freedom by having their shores frozen in concrete. The wildlife, that once thrived in abundance, has all but disappeared and its existence now survives only as a memory in books and songs.

Thus, as soon as my plans to emigrate from Austria to Canada were finalized, I started planning a special project: to find and observe as many of the North American wild animals as I had seen pictured in books.

THE BEAVER

North America's largest rodent, the beaver was only known to me from books and photographs. It was not until after I had arrived in Canada that I saw a beaver in its natural surroundings for the first time. I had read Grey Owl's book "The Adventures of Sajo and her Beaver People" with much interest and delight and often dreamed of a little cabin in the wilderness near a beaver pond before I settled in Canada.

Eventually I saw myself live the rugged life of an outdoorsman deep in the north of Canada.

I started my new life in a small town in the middle of the Ontario bushland, where I gradually learned the difference between reality and daydreaming.

Months after my arrival the company I worked for closed for vacation and my landlord suggested that I spend my first Canadian holiday alone in his remote wilderness cabin so that I could see whether I would really like to live in the wild by myself.

The cabin, sheltered by trees, stood on the outer edge of a little bay near a lake called Tomiko. A path ran down to the handmade dock. The living room offered the tranquil view of sky, water and bush.

As my first day wore on, and the silence grew, I thought, "If this is an example of life in the wild, I am for it." I was elated by the grandeur of the evening and felt as happy as a well-fed cuddled baby. With that feeling I fell asleep.

I awoke a few hours later to a horrible noise at the door. I had the distinct feeling that I was in immediate danger.

The room was bathed in the ghostly light of a full moon that seemed too bright for my sleep drenched eyes.

Wild animals are trying to break into the cabin! Wolves? No, it must be bears! Thoughts ran wildly through my confused mind.

I jumped out of bed, grabbed an axe and braced myself for the last stand. I had sudden flashes of my mutilated body lying on the floor.

The onslaught I awaited on with palpitations did not happen, yet the noise that had frightened me into my stance of self defence, persisted. I realized then that it came from quite a distance away. Peeking through the window I saw six beavers busily cutting down trees. The quietness of the night had magnified

the sound of their gnawing and my lack of wilderness experience had let my imagination run amuck.

The beavers kept up their work until a new day crept over the horizon. Then one after another they went down to the water and like nymphs out of a fairy tale disappeared, leaving behind an assortment of felled trees and chunks of wood.

Although I searched almost all day along the shore I did not see them nor could I find a dam or lodge.

When evening came I went for a swim and later sat on the dock to let the sinking sun dry my body. As I glanced about I noticed ripples on the water's still

surface. It was a beaver heading in my direction. When he was about ten feet away he noticed me. He kept his eyes on me as he slowly continued his swim.

His ears perked and his nose quivered to size me up. Then he left my field of vision. I remained motionless as I did not want to scare him away. The next few seconds brought a complete reversal of intent between the animal and me. I became suddenly frightened by the explosive shot that cracked unexpectedly through the quiet evening. The beaver's broad tail had slapped the water's surface as soon as he had caught my scent. This warns other beavers of

18

possible danger, or scares a predator into flight as he makes his escape under water.

Transparent eyelids protect his eyes under water. Ears, nose and mouth become instantly watertight. The webbed feet of his hind legs provide propulsion, and the tail serves as a rudder. Due to changes in blood pressure, heartbeat and blood distribution, the beaver can easily stay under water for several minutes without having his swimming ability impaired.

This one popped up after two and three-quarter minutes, which was long enough to make me believe he had left for good.

He repeated his performance a number of times before he accepted my presence, realizing that I meant no harm.

Beavers are sociable creatures and soon others arrived. Perhaps the second one did not trust the forerunner's judgement because he too slapped his tail. After that mutual trust was established and soon six of them were busy eating and felling trees as they had been doing the night before.

Before the white man's arrival in Canada, Indians had treated the beaver with due respect, but at the turn of the century the white man's greed for fur almost exterminated this rodent. Some Indian tribes credited the beaver in one of their legends with the creation of land at a time when the earth was completely covered with water. At that time only

those animals existed that could live in the sea: fish, turtle, salamander, muskrat, and beaver. Then, miraculously, a woman fell from the sky who was to become the mother of mankind. But people need land to walk on and so the animals tried to bring some earth from the bottom of the sea. One after the other they tried and failed, except the beaver. After having stayed underwater for such a long time that no one hoped to see him again, he reappeared, almost lifeless, clutching a bit of mud in his paws. From this the sky woman made the land with its valleys and mountains and later gave birth to the first two brothers who became the fathers of all tribes.

The beaver is a very industrious animal and he works hard for good reasons. Since his four incisors

20

never stop growing he must chew wood in order to grind them down, or else they would grow too long and he would no longer be able to feed himself.

The beaver feeds on grass, shrubs, berries, weeds, rushes, water plants, roots, and the bark of trees such as maple, birch, aspen, dogwood, and willows.

If he is not eating or cutting down trees one might find him at work on his dam or his lodge. He builds the dam to flood the entrance of his lodge to keep out predators, and sometimes to raise the water level of the lake or pond. This makes it easier to get to higher standing trees. In winter he is seen much less than during the rest of the year. He feeds from his underwater stockpile and though he still travels a lot, he stays mostly under the ice, where he gets the necessary oxygen from air trapped in pockets.

The highlight of the winter is mating time. Four months later up to six young ones may arrive, fully clothed in fur and with their eyes open. Hours after birth, they are ready for their first swim. For the next two years the family remains together, but then the young generation is chased away to make room for new arrivals. Thus the beaver makes sure that the area he lives in does not become overpopulated.

On many other occasions I have been able to observe beavers during my wilderness boat trips in Ontario and other parts of Canada. I have always found them friendly and trusting, except for one time.

Years later a friend and I spent a few days in the bush on a kayak tour. Because of the various water levels of the different connecting rivers, we often had to backpack our gear and kayaks. It was on one of these portages that we ran into a little beaver who behaved strangely.

fore foot

We were both carrying our kayaks in similar fashion, on one shoulder and held with one hand, while the other carried the paddle. We followed a worn deer trail that ran along the river. My friend was ahead of me some twenty yards. Suddenly he dropped his kayak and wielded his paddle like a sword, defending himself against the attacks of a beaver. The little animal charged him from many angles and kept him on his toes like a boxer in the ring. I was not sure whether he was afraid of the beaver's teeth or tail, and while I succumbed to the humor of the situation, he was cursing. The beaver, however, perhaps not too organized in his attacks, allowed himself to be caught by a blow that threw him off balance and gave him a new sense of direction: he quickly fled the battle scene and vanished out of sight.

hinde foot

Our inquiries into the unusual behaviour of this beaver brought bewilderment and disbelief. The only credible explanation was that the beaver might have had rabies.

By the standards of a seasoned trapper my landlord's cabin was not a true wilderness retreat.

22

Since I had lived all my life in a crowded little country, I felt as if I were a thousand miles from the nearest person. If an Eskimo were to spend a week in central Australia it would probably have the same effect on him as this one week at lake Tomiko had on me. The contrast was stark. Unfamiliarity bred fear. I felt lost. Uncertain.

On the second day I did not know what to do with myself. I had nothing to read in my own language. English paperback books were lying around in the cabin but my mind was not yet ready for the many

strange words and their meanings.

I thought sketching would take up a lot of time, but after I had sketched the cabin from different angles, the trees, the shore and the little dock, I needed a change of activity. It seemed like a good idea to go for a walk, but the bush offered no trails. The shore, cluttered with rocks and fallen trees, soon proved an impossible choice for a hike. I settled for the path to the road. It was narrow and overgrown but at

23

least good for walking. In Austria, walking was the cheapest form of transportation and therefore my legs were in good shape. Soon I walked the three-quarters of a mile to the road. I thought if I followed the road I would come to a clearing. It was a bright day and I longed to see the surrounding country.

A quarter of a mile down the rough road a creek took delight in turning the ground into twenty feet of happiness for waterspiders. Butterflies sailed like bits of coloured paper through the air. One of them landed on the mud and as I moved closer to examine it, I noticed impressions.

A cow must have walked through here not so long ago. Where there are cows there are farms, and where there are farms there are people. Good! I felt like seeing people.

Then I remembered my landlord telling me that nobody was around during the week since people only came to their cottages on weekends. So it could not be a cow. No cow meant no farms, no farms meant no people. I was alone, deep in the bush. These tracks must have been made by a moose.

Again I remembered my landlord talk about the danger of facing a moose during the rutting season. Was this the rutting season? The realization that I knew nothing of the sex life of wild animals set me off at a fast pace back to the cabin.

To recover from the run I sat down in the sun on a cozy spot and pondered my situation. I did not know how to live in the bush. I did not know its language nor its inhabitants. Everything I knew did not relate to this part of the world. I was a stranger here. From now until my landlord was taking me back I would stay within sight of the cabin.

It was the best place to get on well with my newly acquired companion in my seclusion: fear.

THE CHIPMUNK

The next day I was up early. The first rays of light crept over the horizon and gently teased the sleepy creatures of the woods as I sat by the water's edge to watch the day arrive.

I made up my mind the night before to become friends with the vastness of Canada's wilderness and the new day gave me fresh incentives. I realized that I must acquaint myself slowly with a land that is inhabited by creatures I had seldom seen in pictures, let alone met in real life. Nature is never in a hurry and man can only learn about her on her own terms.

A thick mist clung to the lake and the trees. Dew-drops were catching the early lights. The rising sun gradually thinned the mist and like the image on a photograph in the developing tray, the land appeared slowly, swaying to the rhythm of the first hour of the

day. After a while the sounds of birds grew louder and the shore on the other side of the lake sparkled in the morning sun.

Soon everything was clear, bright, dry and warm. The water beckoned.

I stripped for a swim, but the moment I stepped into the water it stopped beckoning. I managed to splash about a bit and jumped back onto the rock, where the sun quickly made up for the warmth the water lacked. I was refreshed and hungry. With bread, cheese and grapes I settled on my favourite spot just outside the cabin.

A chipmunk played hide-and-seek with himself on the woodpile nearby. Then he noticed my odd breakfast. With quick dashes back and forth, he tested my reactions. Finally, he became bold enough to scurry over my outstretched legs. That established

my harmlessness and made him choose me as his friend. Naturally, with a friend you share your breakfast. He did not think much of the bread, the cheese he liked, but the grapes proved to be his favorite.

As if I were not convinced of his charms already, he assumed a delightful stance while he twisted and turned the grape with his forepaws. His four sharp incisors which are common to all rodents, quickly whittled the grape down to nothing; then he looked for more. Another grape disappeared into the depths of his stomach before he was overcome by the desire to

27

store some away for lazy days. The next four grapes were quickly placed into his cheek pouches, the fifth one he held between his teeth and then made off, streaking to the woodpile where he vanished. A few minutes later, he was back for more. Between the two of us we finished the grapes, that is, he ended up having stored away some at home, while I had eaten all of mine. Still he kept coming for more. All along I had hardly moved so I would not scare him away; now

I had to get up to see what else I could share with him. I moved slowly and he showed no fear. I left the cabin door wide open and just as I cut up a coconut he came in to perform a little tap dance on the masonite floor.

The new handout reminded him of his duty and with full cheeks he made off again. When he reappeared he stayed by the woodpile and ran back and forth, obviously somewhat disturbed. The reason for his agitation soon became apparent. He tried to entice his mate to follow him. Eventually, with his constant encouragement, she left the safety of her hideout. He ran over to me and stuffed his cheeks while sitting next to my hand. The demonstration broke her resistance completely.

During the next days we spent many satisfying hours together. For me the chipmunks were good company and for them I was a good provider. I gave them the run of the cabin which sometimes produced comic situations, when they tried to chase each other on the smooth floor. Once, one of them threw himself into high gear to dash off with a big chunk of coconut that had replaced his centre of gravity, resulting in somersaults, flips and staggering.

A chipmunk always seems to be a little tease and perhaps there was good reason for Grizzly Bear to be angry with him. According to an Indian legend the chipmunk originally had no stripes, but since he kept teasing the big bear he invoked his anger. Chipmunk was not quick enough as Grizzly reached out and the huge claws left scratches on the little fellow's back. Ever since he has borne five black and four white stipes as a reminder never to tease Grizzly again.

To the chipmunk the ground is a huge supermarket and he makes good use of his built-in shopping bags, always collecting food and storing it. Any daily surplus goes either into his winter den or in

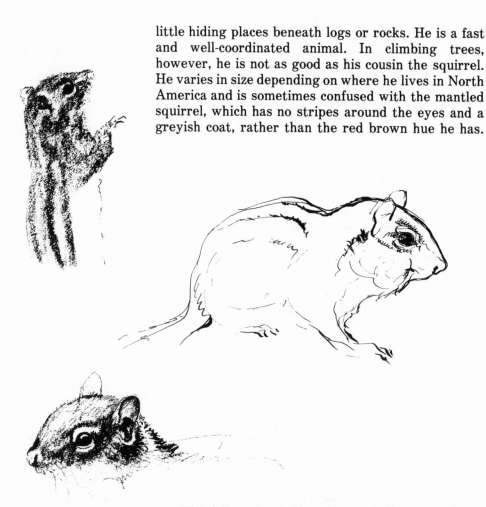

little hiding places beneath logs or rocks. He is a fast and well-coordinated animal. In climbing trees, however, he is not as good as his cousin the squirrel. He varies in size depending on where he lives in North America and is sometimes confused with the mantled squirrel, which has no stripes around the eyes and a greyish coat, rather than the red brown hue he has.

The fall is the busiest time for him. He has to collect enormous amounts of edibles for his wintering burrow which he keeps filled with goodies for the times he awakes from his winter sleep. Sometimes he will even venture above ground in the winter, but only for short brisk exercises.

One can usually find him at campsites and in rural communities where the houses often border the bush. There, people will literally find him on their doorsteps. Many motorists will glimpse him as he often scurries across the highways with his tail high in the air, like a finger on legs.

30

THE WEASEL

Much of my time I spent at the lake's edge, often sitting on the boat dock. Saturday had arrived despite the slow moving days. I looked forward to seeing people again. Strangely, as I looked at the bay and the cabin I felt a familiarity for the place that made my stay appear in a pleasant light. My European past

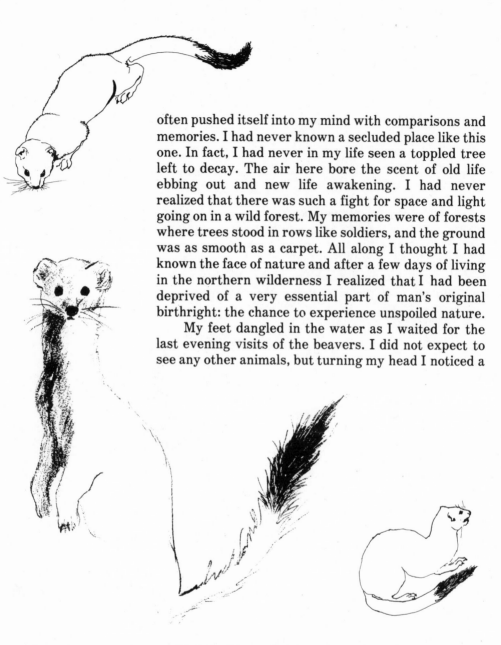

often pushed itself into my mind with comparisons and memories. I had never known a secluded place like this one. In fact, I had never in my life seen a toppled tree left to decay. The air here bore the scent of old life ebbing out and new life awakening. I had never realized that there was such a fight for space and light going on in a wild forest. My memories were of forests where trees stood in rows like soldiers, and the ground was as smooth as a carpet. All along I thought I had known the face of nature and after a few days of living in the northern wilderness I realized that I had been deprived of a very essential part of man's original birthright: the chance to experience unspoiled nature.

My feet dangled in the water as I waited for the last evening visits of the beavers. I did not expect to see any other animals, but turning my head I noticed a

small dark figure darting along the sand, just a few feet away from the dock. A dark-brown slim body, short legs, swift harmonious movements, strong black eyes, little round ears, a long furry tail. Now there was someone I recognized! As a child I had seen a fellow like him raise fury among our chickens. Then it leapt

32

on to the dock with bared canines and hissed. There I was, sitting on a tiny dock in the middle of nowhere without a stitch to wear, with a ferocious weasel two feet from my navel, seemingly deciding on how to get at my throat. I felt as vulnerable as a goat tied to a stake in a tiger trap. The Austrian Gypsies believe a weasel brings bad luck, and at that moment I instantaneously shared their superstition.

"Just leave me alone!" I yelled, with a streak of hysteria in my voice. Uncertainty showed in the weasel's aggressive eyes as he backed up. Continuously hissing threats, he paced the end of the dock. If I jumped into the water, would he follow? No doubt he

could swim faster than I.   When he leapt closer for what seemed an all-out attack I yelled and splashed water over him. This brought an end to the ridiculous situation, for he retreated into the bush.

The members of the weasel family are as different in size as they are different in their choices of places to live. They can be found in wooded areas or in the prairies and range in size from a mouse to a full-grown fox. Best known of all the weasels for his fur is the mink. Best known for his ferocity is the wolverine. In colder regions the weasel will exchange his brown summer coat for a pure white winter growth. Then he gets the name ermine. His smooth white fur has for centuries been of special significance to royalty everywhere, which made him a much trapped animal.

The whole weasel family is known for their fearless hunting habits and they exercise little discrimination in their choice of prey. The wolverine has been known to attack bears and mountain lions. The courage with which the weasel closes on his victim is only matched by his speed. An animal many times

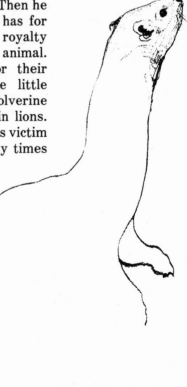

his size will flee in panic, when attacked. Usually the weasel will aim to bite through the muscles of the neck just below the head. If a rabbit is selected for dinner the weasel will jump in its back to administer his deadly grip. The strength of his jaws and the sharpness of his teeth will make it almost impossible for the frightened victim to shake him off. His sharp claws allow him to extend his hunting into forest regions, where he preys on birds and squirrels. His appetite for meat is so great that his young will feed on it before their eyes are opened.

Because they are almost devoid of fear, or perhaps lack the ability to distinguish sizes, they will attack dogs, even man, expecially if they think themselves cornered.

In Indian folklore there is a story that demonstrates the weasel's fearless fighting spirit quite well. A superior natural being who was the benefactor of the Indian race and a friend of all earthly creatures was once captured by another supernatural being, a cannibal-monster. This monster was to devour the benefactor and just as he was preparing for the feast, a weasel happened to come by. Recognizing the predicament of his friend, the weasel acted so quickly that the monster's eyes could not follow his movements. In the flash of a moment he was in the cannibal's mouth, had run down his throat and had eaten his heart. Before the monster toppled over dead, the weasel jumped out of his mouth again. The benefactor decided that such a deed should be rewarded with an eternal display of fearlessness and speed.

THE MALLARD DUCK

In every newcomer's lot is the need for flexibility, mostly of an involuntary kind. The first few months gave me the opportunity to exert it as I rushed through a number of jobs. This "here today, gone tomorrow" work situation did nothing to strengthen my ego, but a far greater problem posed by the culture shock that

every person, who comes from a non-English speaking country, experiences. Because of my inability to communicate fluently in English I was very grateful when a young German immigrant offered me a job in his gardening and landscaping business. It was a good way to get close to the country that was so new and strange to me. As I planted shrubs, dropped seeds into furrows, weeded flower beds, trimmed plants, and touched the soil I felt a kinship of belonging to this earth even if I did not yet belong to Canada. My new boss drove me to each customer's place and left me there with tools and instructions. Though I hardly spoke to anyone during the day my English improved as I continued to teach myself. Each evening I wrote new words and sentences on a piece of paper and the next day, while at work I engaged in make-believe conversations.

The gardens I worked on seemed to belong to very rich people. The residence of one of the customers was at nearby Trout Lake. The elaborate house and garden were most impressive. So was the seaplane docked next to the huge power boat, the tennis court, and all the large cars, that suggested constant guests. What fascinated me even more was the lake. Surrounded by gentle hills it stretched through a mixed forest that brought the wilderness right into the rooms of those people who lived in cottages along the shore. Like all unspoiled lakes it bore the mystery of life-giving forces beneath its reflections and sent out generous invitations to man to sit and listen. During each lunchbreak I followed its call and on the water's edge, waited for the revelations, that are the rewards of patient observation.

There were many song birds that I heard for the first time in my life. Water birds I had never seen before swam in coves shadowed by the hills. Trees—unmolested by man—grew along the shore, sometimes tipping their tired crowns right into the water. The undergrowth was thick and wild with plants and shrubs.

How different from the forest in Austria, where one soon had to fight boredom in seeing rows and rows

38

of planted trees. What a contrast to the freedom nature displayed in Canada where everything teemed, where plants and trees throbbed with life and crowded each other in their struggle for light and space. This was a country where one could obtain happiness through the wholeness of nature. A happiness expressed by wild animals whose sentiment for this earth is neither as remote nor as abstract as science wants us to believe. One has only to sit by a lake and watch in silence to understand.

Small waves gently sloshed the rocks and a cluster of grass swayed with a similar rhythm. The trees whispered. Close-by but hidden from view ducks chattered mildly. A flock of black-capped chickadees settled in a nearby bush. None seemed to be first, none last. They lived in the harmony of companionship. Politely they inspected each blade, pointed out tidbits to each other and exchanged pleasantries while they shared their meals. They came and went together.

Then Mallard ducks drifted into view. Bobbing on the waves they rounded a huge slab of granite that a remote groaning of the earth had placed there. Ahead swam the female and, as if holding on to an invisible

string, nine chicks followed—one behind the other.

Trout Lake is a small body of water that spreads in an east-westerly direction for some 15 miles and harbors many species of waterbirds, the mallards being residents only of its small cattail marshes that mix with tiny mud flats. Though a sudden wind can

make the water unsafe for boating the shores never rebound with the crash of giant waves, making the lake's numerous coves ideal places for mallard parents to rear their chicks.

The mallards are reputed to mate for life and build their nest on dry land in grass or in shrubs, mostly close to water so that the chicks, soon after hatching, can be led to it to literally demonstrate "how ducks take to the water".

The chicks were covered with pale yellow down and wore darkbrown "hats". Obviously delighted with their ability to swim so deftly they found it hard to refrain from exploring and often obeyed their mother's frequent quack-quack reluctantly. The hen wore a dark brown plumage that was tinted with grey. The males had left the females with the responsibility of incubating the 10—12 eggs and rearing the young.

40

The drake, like most male ducks, proudly displays the colorful outfit. The feathers on his head and neck are of a deep blue-green color, the bill is yellow and a narrow ribbon of white circles his neck. The brown wings have purple patches that during a fight show up as brightly as during rest time. After the breeding season the drake will lose his colors and his flight feathers and with them the ability to fly. During the moulting period the drake resembles the hen and they rely on their concealing plumage for safety until the sky becomes their domain once again.

The hen swam towards the shore quacking loudly and the chicks followed puddling wildly. When the hen's feet touched ground she stood up, lifted her body out of the water and flapped her wings, as if to show the chicks what they must do to become airborne. Several of them ignored their mother but some tried to imitate her and failed miserably, mainly because of their undeveloped wings. Then the hen led the whole group into the thicket where a trail ran to higher grounds. During the day they would feed in open country or in fields, where they would look for grain,

vegetables, insects, slugs, and snails. The nights and mornings the mallards always spend in creeks or marshes. On land they are very sure-footed, because their feet are not as far back as those of other waterbirds, such as the redhead duck, the ringnecked duck, or the canvasback.

Mallards' fondness for small ponds goes even as far as taking over suburban swimming pools or decorative fountains, as I could observe years later in Vancouver. There, in front of a 26—storey highrise building, oblivious to the downtown traffic and the steady stream of people coming and going, a mallard couple had chosen a bubbly fountain as their summer residence. Like a pair of confused tourists they would spend the day standing huddled together, accepting handouts from kind-hearted residents and displaying courteous affection for one another. The drake, though more aggressive in general, would always stand back and wait till the hen had fed before he too would take the tidbits the city people offered.

It takes a good deal of rough weather to send the mallards south and if there is somewhere unfrozen water they will stay in it until the cold practically forces them to look for a friendlier climate elsewhere.

During the warmer season the mallards prefer

the shallow lakes, rivers and marshes of the northern
wilderness, and in farm areas they will quickly take
over any pond that allows them to reach food by
submerging their heads under water to dabble in the
muddy bottom. Because of this characteristic
behaviour the mallards, together with all other ducks
that feed in the same manner, belong to the group
known as dabblers.

Ducks that stay fully submerged for several
minutes when searching for food are called divers.
These divers have smaller wings and prefer deep
water. When approached they will run on the surface
of the water to get airborne, while the dabblers are
capable of lifting themselves almost straight out of the
water. Though it is quite obvious that the much larger
wings of the dabblers help them to take to the air so
quickly, the Cree Indians have a different explanation
for it. They refer to one of their legends that explains
how they think the mallards were given such a quick
get-away mechanism.

A young girl and her lover were pursued by an
evil shaman who wanted the girl for his wife.
Cornered, they saw themselves doomed to a horrible
death but were saved by a kind spirit that changed
them into ducks seconds before the evil shaman

43

arrived. Just as he reached for them they leapt straight up and flew away. Ever since, the mallards have been able to take to the air almost vertically.

In time Trout Lake became a source of many happy and rewarding hours, not only because I was able to observe water birds but also because of the many hospitable people who opened the doors of their homes and cottages to me in friendship. Most memorable of all was a softspoken grey-haired bachelor called Bill whose cottage was a regular visiting spot for young people on weekends. There was plenty of gasoline for the motorboat, a canoe, water

skis, a diving board, a roomy cabin, a fridge filled to the brim with food and soft drinks, and the kind-faced Bill, making sure his visitors had a pleasant time. With great relish he would tell stories in the evenings of the time when he was a young boy - when horse and buggy were the popular mode of transportation, when streets were unpaved and gentlemen stepped off the wooden sidewalks into the mud to make room for ladies. While we were huddled around the fire the flickering shadows of the trees created the mood of eeriness. We listened to him and the moaning owls and could almost feel the closeness of the monster that he thought lived in the depths of the lake.

44

THE DEER

Each day brought new fascinating experiences that often bewildered me beyond concealment. I had never before lived in a country where one could walk the streets of a medium sized town and find the wilderness begin where the streets ended. I had never seen deer forage flower beds and vegetable gardens, racoons climb roofs, black bears overturn garbage cans, and arctic owls stare from laundry posts. There was an unlimited number of fascinating things for me to discover.

Late one evening in the middle of September the northern lights illuminated the sky with a pale hue as I walked home. The first two hours it changed little; sometimes it only glowed like a far away forest fire. Then as the paleness ebbed from the sky, shafts of green light crossed flashes of red and blue, changed to

ribbons that drifted into veils of colored light that
seemed to be dimmed and brightened by invisible
stagehands. At times the color seemed to drip from
the sky, then to flow across the horizon. Finally at 3
a.m. the invisible switch was turned off and the stars
returned. Tired, but excited, I went to bed.

Next day at the shop I talked to Denton, (with
whom I worked most of the time), about the spectacle
and he explained to me that what I had seen was less
brilliant because of the city's bright lights. To really
see northern lights one has to be out in the country,
far away from any city. The hunting season was
coming up and he and his friend Mike had planned a
few days in the bush. I was invited to come along and
perhaps see another sky spectacle. If there were no
northern lights I could surely see stars, brighter than I
had ever seen them before. Besides, they might even
get a deer, which in itself was another Canadian
experience for me.

One Friday evening two weeks later I rushed
home from work for a quick snack, changed into my
warmest clothes and with the parcel of food my
landlady had made up for me, found myself bumped
around in the back seat of an old car that rattled up a
country road. Denton and Mike were chatting away in
the front, still unsure where the best place for deer
would be, while the heater spewed out warmth that
made me drowsy. Hours later, the car came to a stop
and Denton declared that we had reached the place
where we would spend the night. He and Mike slipped
into their sleeping bags to ward off the cold that
flooded through glass and metal as soon as the engine
was shut off. Sleeping bag? That must have been one
of the many words amongst their instructions that I
had not understood. The closest thing to a sleeping
bag that I could come up with was my woollen scarf. I
wrapped it around my waist under my sweater and
drew my legs up to stay as warm as possible. This was
supposed to be my blackest night with the brightest
stars. Instead it turned into the coldest night. "Oh
well," observed Denton dryly while he zipped up his
bag from the inside, "This is Canada. You want one

46

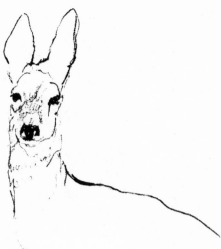

thing and you get another."

Some Cree Indians believe that during the night a man on his own should not be without a fire. The night has strange powers that take possession of a person's mind. A fire may be the only protection against any harm that lurks behind the trees and rocks. Spending the night in the bush without a fire did not seem to bother Denton or Mike and certainly did not keep them from snoring. The more sleep tried to overcome me the more the cold fought to keep me awake until eventually this tug-of-war took on strange forms. While I shivered I imagined myself sitting by a roaring fire until my senses lost the feeling for reality and I was helplessly given to fantasies that live in the moments before sound sleep comes. I heard trees speak to each other and saw the face of the night change into a timeless existence when all animals were people. They drifted in and out of view, all shivering, miserable, stiff and sad. Some froze if they stopped to rest while boulders split wide open and branches broke off trees like icicles. It was as if the earth was cast into the far abyss of the universe where only the darkness and the cold live triumphantly. Finally I fell into a deep sleep.

I was awakened as someone shook my arm. It was Denton who with a friendly grin offered me a steaming cup of coffee. I saw Mike break eggs over the frying pan that stood on the hood of the car. When I smelled the delicious bacon that sizzled invitingly I felt the warmth return to the marrow of my bones. The last bit of stiffness was forced out of my body by a brisk

walk, as we went looking for deer tracks. While
Denton and Mike crunched through the snow I
trudged behind them wondering how they would ever
get near a deer making so much noise. Occasionally
one of them motioned to stop and listen. There was no
sign of a deer anywhere. The trees stood motionless. It
was a quiet morning, most of which we spent walking
and listening until it seemed necessary to move on and
so the car once again took us away.

After hours of driving on backroads we turned
into the bush where a slightly overgrown road bare of
snow led us to a stand of birches and maples.
However, while the snow was gone, a heavy layer of
leaves covered the ground and gave away every
movement. As though the message was not loud
enough already, a squirrel called out: "Watch out
everybody, here come some people." We walked along
the abandoned road until Denton decided that this was
not getting them anywhere either. "You stay here
Helmut. Mike, you and I will scout the area up there. If
you see one, Helmut, chase it up towards us." With
these vague instructions I was left standing in the
bush, while they disappeared through the trees. I sat

49

down on a log and looked around. Now and then a tree groaned, a leaf wobbled to the ground here and there. It was quiet and peaceful. I was not happy that they had left me alone. What if a bear should come? If I had only borrowed a gun from somebody. I did not even have a knife. Perhaps I could climb a tree if some wild animal would charge. Though the Crees say a man should not be without a fire during the night in the bush, perhaps they really mean that no man should be alone in the bush. I whistled and listened for a reply. There was none. How far did they go? I hoped they would remember where they left me. Just then a wild commotion broke loose somewhere behind me. I whirled around to face the brute that no doubt was charging, but saw nothing. The noise had stopped abruptly and I could only hear the thumping of my heart. Then it started up again, this time closer to my left and to my relief I spotted a chipmunk that scurried through the leaves on the ground.

Well, if a little animal makes so much noise, I reasoned with myself, then surely I should be able to hear a big one, such as a bear or a moose, from a distance. There will be time to run, I figured and relaxed again. I do not know how long I waited for the two hunters to return. At one point I caught the scent of cigarette smoke and a few minutes later they came shuffling down the road. They had spotted a small ravine and that was where the rest of the afternoon was to be spent. Mike sat down near the opening of the ravine. Denton and I posted ourselves at the other end. There we waited in silence.

Before my eyes there was a drama to unroll that was to teach me that luck and misfortune are based on the same event.

We heard something approach from the left and before I realized what had happened Denton jumped up, aimed and pulled the trigger. A numbness went through my ears as the shot bellowed and rolled off a distant hill. On the bottom of the ravine a white-tailed deer limped through the tall grass hit badly, but desperately trying to outrun her fate. I jumped up excitedly just as Denton fired again only to let myself fall to the ground, overcome by uncontrollable fright that the nearness of the high-powered rifle shot sent through my limbs. It was the first time in my life that a gun was fired only inches from my ears.

Below in the ravine the doe's front legs buckled under her while she made running motions with her rear legs. She tried to get up again but her strength left her with the flow of blood that came through the hole in her chest. She sank to the ground and shook as if to ward off the chilled hands of death. Again and again she tried to get up. "For heaven's sake, kill her!" I yelled at Denton. Another shot at close range severed the ties that bound life to this animal and with a groan her head sank down to rest on the grass.

I knelt beside her and felt her warm body, as a great sadness surged inside me.

Mike came running down and broke the spell of tragedy with jubilant shouts. Denton joined in hollering his joy over the lucky hunt.

51

Lucky indeed they were. Something had alarmed the deer and driven her into the ravine running with the wind instead of against it. Later on, when her milk glands were found to be full, the reason for her panic became obvious. She had tried to lure a predator away from her fawn by keeping her scent in the wind. Normally she would have been able to spot us. Her eyes though unable to focus sharply into the distance enabled her to detect a flicker of movement. Her hearing was acute and she knew the sounds of nature and the sounds of danger. Her sense of smell was the keenest of all. With it she analysed and identified even the faintest traces that the wind carried from miles away.

Somewhere in the bush a fawn or two were waiting for her return.

The doe lay on her side paritally exposing her belly that was covered with white fur. The white chest was crusted with blood and her white throat patch was similarly stained. On the back and on the flanks her coat was greyish with a hue of brown. This was the color of the thick winter coat, that would have kept her warm and cozy even while she slept in the snow during the night. Next spring she would have exchanged it for a thinner summer coat of reddish brown.

The front legs of the dead animal were hidden under her chest, the rear legs were stretched out exposing part of the white insides. Halfway on the outside of the lower section of her top leg was a small lump. It was the tarsal scent gland, one of three kinds of scent glands common to all deer. These glands are used to mark and signal. If a fawn is to be warned of danger a doe will discharge her tarsal gland. To recognize her own kind or to retrace her way through unfamiliar terrain she will use the scent that is given out by the gland between her toes.

I looked at her bushy tail that was still signalling alarm. It was pointing away from her body exposing the white underside.

In Indian history this tail was once most important. Because of it we can now cook our food and

52

warm our houses. They say the world was not always as it is now.

Long ago the earth was dark and cold. There was a Great Evil Spirit in the sky who jealously guarded the sun, the moon, the stars and fire. The animal people lived in misery and decided to steal the fire. One after the other they tried and failed. Only Deer remained. At that time she had a long thick tail. In graceful silent leaps she disappeared into the sky. When she was visible again some time later she carried a flame on her tail trailing sparks through the sky. The earthlings quickly lit their torches but in the meantime the deer's tail had burnt to a short stump. Ever since, all deer have short tails. Some, like the black tailed deer or the mule deer display even to this day bits of singed hair on their tails.

The whitetailed deer is the most abundant of all

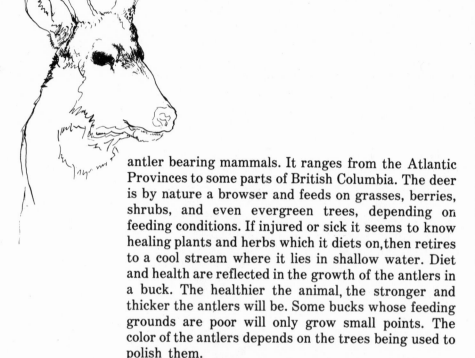

antler bearing mammals. It ranges from the Atlantic
Provinces to some parts of British Columbia. The deer
is by nature a browser and feeds on grasses, berries,
shrubs, and even evergreen trees, depending on
feeding conditions. If injured or sick it seems to know
healing plants and herbs which it diets on, then retires
to a cool stream where it lies in shallow water. Diet
and health are reflected in the growth of the antlers in
a buck. The healthier the animal, the stronger and
thicker the antlers will be. Some bucks whose feeding
grounds are poor will only grow small points. The
color of the antlers depends on the trees being used to
polish them.

A young buck develops spiky antlers in his second
year and they in turn will fork out in his third year.
After the fourth winter they will reach their basic
structure with three points on either side. After that
they may grow one or two more points but seldom
more than five on each side.

The antlers are actually fast growing bones and
surely belong to the wonders of animal life.

A full grown buck will grow a set of impressive
antlers within a few months every year. During the
growing period a skin, or velvet covers the antlers. In
late summer the blood vessels and nerves in the velvet
cease to function and dry up. The buck then sheds the
velvet by rubbing the antlers against trees. By the
middle of November the mating season has arrived

and the antlers are ready to be used in combat against a rival in search of does.

Ironically, the same antlers that portray life and vitality also invite death. The best looking bucks are sought after by hunters who in their confusion of life's true values like to see a trophy hanging on their wall rather than grant the animal the right to perpetuate its species. In this way they contribute to the downbreeding of the deer. For every prime buck shot a somewhat inferior male is allowed to breed. Natural predators, such as the wolf, the lynx, the bear, or the cougar display more concern for nature by weeding out the weak and diseased.

The spirit of nature comes to life when in the early hours of a November day a buck trumpets his challenge through the mist. I like to think that the sight of two bucks fighting for reasons that are as powerful as the pull of gravity will teach any man more in one hour than a hundred teachers can teach him in a lifetime. I know I am wrong. Once a man who had outgrown his fascination for nature came upon two bulls engaged in a fight. The hunter shot the larger one which went down locking his antlers with his opponent's, who tried to get away. The hunter lost no time killing the second animal and now proudly displays in his mansion the mounted heads of the two bucks with their antlers still locked.

In some alpine valleys in Austria the same kind of greed caused the death of many significant bucks in the last century. They were shot by poachers because the lower section of the antlers fetched a high price when brought on the market as a powder. It was secretly sold in small portions as an aphrodisiac, though most of these remedies stimulate the imagination of men more than they improve their virility.

It seems that all deer like the proximity of cultivated land and often visit farms and cattle ranges. They are also attracted to highways where they find the much needed salt. Even during the summer they will congregate near curves, where the rain has washed the salt into the lower areas. Many serious accidents have happened to people who made no allowance, while driving on Canada's highways, for the deer's craving for salt.

Death comes in many ways to the white-tailed deer. The ageing process is closely related to the condition of the molars in each animal. In old age the molars are usually ground down to the roots and prevent the animal from getting the full benefit of the food he tries to chew.

Man, and man's best friend, the dog, are competing for first place in decimating the deer population. The pros and cons of hunting are continuously subjects of heated discussions while the dog's deer killing activity goes on relatively unnoticed. This is mainly because rural communities allow the dog great roaming freedom. In nearby wilderness areas dogs will quickly band together to hunt. The scent of a deer will often turn the friendliest dog into a killer. In less than one hour a dog can run a deer down, kill it and return home to the unsuspecting owner. The problem lies less with the dog than with the dog owner, who cannot imagine that his pet, such a charming companion to his child, turns into a vicious killer at the sight of a deer.

At the time when the earth was young the animal people lived everywhere and infringed on each other's rights. The Creator-God decided to end their confusion by giving each animal a special place to live. He summoned them to a meeting. All arrived except the dog that was nowhere in sight. Finally, it was decided not to wait any longer and the Creator—God proceeded to designate habitats to the animal people. The water he gave to the fish, the air to the winged creatures; four-legged animals were to live on the ground, others below. Finally when all the places were given out the dog arrived. Various animals suggested that the dog should live with the wolf or the coyote but the dog wanted his own place. The animal people were eager to go home and to bring the meeting to an end the deer proposed that dog should live with man. All agreed quickly and the Creator—God commanded the dog to share man's place. Dog's anger was so great that he promised never to leave Deer in peace and to chase it whenever he could.

56

THE RED FOX

A common folk tale has it that all Austrians are born on skis, drink wine instead of water and yodel. With one short demonstration I usually convinced any listener to abstain from further requests and also established my new reputation: I was an odd Austrian who could not yodel, did not like wine, and worst of all, could not ski.

Unfortunately I had arrived in Canada at a time when Austria's fame as a nation of downhill ski champions was still at its peak. I saw in skis only the means to get out during the long winter months and explore the countryside. I felt a simple wooden pair of skis and inexpensive boots were good enough, but a salesman in the sports store was not in agreement with me. As a young Austrian immigrant with the figure of a ski instructor, I should have the most

57

sophisticated equipment. I tried my best to make him understand why some Austrians are downhill racers and some are only beginners, but he remained adamant. Finally, convinced that I was to blame for the lack of communication, I resigned myself to going home without skis. Just as I was about to exit I noticed some strange looking frames with leather webbing. I made some inquiries and a few minutes later left the store with snowshoes instead of skis.

In the privacy of my room I practiced the first steps on them. While I thumped back and forth I recollected what I had been told about them.

This particular pair had frames that were made of solid ash and the thongs were cut from raw moose hide. Today when you go into a store you may find snowshoes that are made entirely of plastic. The National Museum of Canada exhibits snowshoes that are made of branches and bark strips and it is believed that the invention of snowshoes dates back several thousand years. Indians used them long before the arrival of Europeans. White people in turn used them in opening up the country when all other forms of travel were impossible. To this day they remain the only means of walking on the snow.

Thinking of Indians and pioneers fired my imagination, and I wondered what my friends and relatives would say if I sent them pictures that showed me tracking through the bush on snowshoes. I had anticipated a picture-taking session and grown a beard that was now six weeks old—long enough to be photographed. For my first trip I decided to visit the islands on Lake Nipissing. I could see them from my window. Earlier, local Indians avoided these islands

which they believed had mysterious powers. A geologist gave a more practical reason. He believed that their shape indicated an extinct volcano that may have given off deadly fumes for a long time after it had ceased to be active.

While my landlord shovelled the driveway clear to get the car freed for the Sunday morning church visit, I strapped on my snowshoes to take part in a different kind of sermon – the sermon only nature gives during an Ontario midwinter day.

The thermometer showed thirty-two degrees below zero on the Celsius scale, the snow sparkled like crushed glass and there was not a cloud in the sky. The snow covered the shore in huge waves and coming down one of the drifts I lost my footing and took my first tumble. Further on where the snow was swept clean off the ice I passed two men who had chopped holes through the almost two-foot (60cm) thick crust to fish. On windpolished ice, snowshoes are a hindrance so I took them off and went on, slipping and sliding, while a brisk wind from the east tried to peel the skin off my face, freezing my breath to my beard.

The three islands were grouped close together forming a broken circle that acted as a wind breaker. In some places, the snow was so deep, I sank into it up to my hips. Again I fastened the snowshoes to my feet. By now I could walk on them without bruising my ankles.

The islands differed only in size from each other, but not in appearance. They had the same scraggly-looking birch stand that was dotted with weatherbeaten firs and an assortment of rocks and boulders.

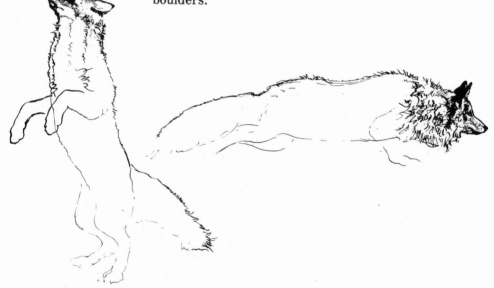

I set up the tripod, carefully selecting a place that would make an authentic background for my element-defying appearance. Then I snapped a series of photographs, using the self-timer, to convey to my folks back home that the ice age was alive and well in Canada.

Satisfied, I walked on to the next island and came across a line of tracks that looked like they had been made by a dog. There were no other tracks and it seemed unlikely to me that a dog would cross over the ice for miles without a good reason. The next animal that came to my mind as having a reason for being there was the wolf. I had heard that a wolf's feet were larger than a dog's. The more I looked at the tracks the more they grew, until I was sure they were made by a wolf.

A fantastic idea came to me. If I could get close enough to take a picture of a wolf and send it home with all the other snap shots, it would take them days to shed the goose pimples.

Silently I followed the tracks that led straight into the wind. I kept one hand on the camera that hung from my neck and the other held the tripod, ready to use it as a weapon. The dotted line led me to the next island, where it wandered about in an irregular pattern before turning back. If I followed it the wind would give me away, so I turned around until I reached the rock-strewn shore. There I turned south and walked along the shore. Stalking an animal is a

slow and exhausting activity, but my patience paid off. Straight ahead was a boulder that protruded through the snow. In a cavity on the sunny side of the rock an animal was gathering in the sunshine. With the patience befitting an aboriginal hunter I moved closer inch by inch, until I clearly saw that the wolf was a red fox.

His relaxed ears and shallow breathing told me that he was sleeping soundly. Protected from the wind, he was making the best use of the midday sun, lying snugly curled up. His bushy tail was draped alongside his belly, the white tip hidden under his chin. His coat glistened in all shades of deep rust and pale gold, accentuated by the colors surrounding him: the greyness of the rock, the blueness of the sky and the whiteness of the snow. The rugged scenery had a peacefulness that made my heart glow with happiness. I watched the animal breathe softly and it seemed that his lips were given to the melodic recital of a universal creed: Nature is the strongest link with God.

I could have stood there for hours without knowing it, but the shadows crept slowly across the snow and whispered of time going by. Finally, the cold delivered a forceful message: you have been standing here too long, get moving! Before I obeyed I saw the fox's ears twitch and straighten up. Just as quickly, he was awake and looking around. Though his eyes were quick to detect any kind of movement he relied more on his highly sensitive ears. He looked at me obviously trying to find out what this shape was ahead of him. All of a sudden he was up and out of sight. Moments later I saw him turning into the trees before he completely disappeared. I was so absorbed I had forgotten to take a picture.

The red fox is a roamer who prefers to do much of his travelling during the night. His eyes are equipped with mirrors that are utilizing every available glimmer of light. Unlike the other members of the dog family, the pupils of the fox are oval, which narrow to a slit in bright light and open up fully during darkness enabling him to see much better than the dog, the wolf or the coyote. Despite his large appearance, which he

gets from the long hair of his coat, the fox is no heavier than a house cat. His diet consists of whatever the season has to offer and all the leftovers he can get, such as a deer, brought down by wolves or wounded by hunters. In the winter he cleans up deer that have died of starvation. Mice, moles, rabbits, insects make up the mainfare of his diet. He will not pass a berry bush without dining on its fruits and is extremely fond of strawberries. His wilderness habitat is near lakes, streams, wild meadows and he is a resident of farmlands as well. There he often causes trouble amongst the poultry.

In May or June four or more puppies can be seen frolicking in front of their den. At that time they may be three or four weeks old. The playful nature of the fox often drives them to play with cows or caribou.

In fables and stories the fox is endowed with almost human qualities, some believable, others obviously attributed to him for reasons of artistry. Many of these stories are read to children around the globe and tell of the fox's cleverness. One Indian legend, however, explains how the fox's curiosity got him into trouble.

At that time all things had voices and   during his wanderings, came to a beautiful lake. He saw some fish in it and prepared to catch them when the lake spoke to him. It asked him to leave the water undisturbed because of the 'Spirits of the Night' that came here to get their food. This aroused the fox's curiosity who wanted to see what those spirits looked like in daylight. He threw many rocks into the water hoping this would force the spirits to appear in

daylight, but when the spirits found the water murky the next night, they were so angered they decided that the culprit should hunt at night to see for himself how difficult it is to find one's food in the darkness. Ever since that time the fox has to hunt during the night.

Several months after I had spotted the fox on the rock I saw another one. This time the setting was out of the ordinary. A newly acquired Indian friend, Gerald Dokis, had taken me to his grand uncle's workshop, just outside North Bay. There Mr. Couchis made his living as a skillful taxidermist. Inside, the shop looked like a cabin in Noah's Ark. Wood ducks and mallards stood or sat next to snowgeese and small birds. Fish with their mouths gaping for food were lying on boards, or peeking from open boxes. Among all the stuffed animals that seemed to have casually gathered, there was also a red fox. For all I knew it may have been the same one that I had surprised during his winter day nap. The fox was frozen into a stance of snarling defiance, his teeth were bared and his ears bent backwards. Placed into the grass outside the workshop it looked so real that I could not resist taking pictures with my old-fashioned camera. One of the photos showed me reaching out with my hand to offer the fox some food. It looked so horrifying to my mother that she promptly begged me in her next letter to come back to civilization.

THE SNOWY OWL

I enjoyed sending home letters and photographs that showed Canada's rugged wilderness, and never missed an opportunity to give the impression that I belonged to the breed of pioneers to whom civilization is forever indebted. Of course the trappers, settlers, pioneers, missionaries, and traders of the past were similarly awed by the vast wilderness and the strange animals they found in Canada. The modern immigrant differs from them in having the comfort of automatically heated houses, electricity, a network of roads, and means of transportation that were beyond the wildest dreams of the men who depended on dogteams, snowshoes and canoes. Though industrialization has made it easy for the modern immigrant to avoid wilderness hardships, the love for the newness still brings about occasions that are lessons in early pioneer day living. This is the time when a newcomer exposes himself as a greenhorn. Webster's dictionary defines the word greenhorn as—'an extremely inexperienced person.' Three such greenhorns wanted to see what the Temagami Winter Carnival was all about.

First there was Gerry, a Dutch sailor whose eyes were always trained on the far horizon. He had left his last ship on the Atlantic coast and was working his way across Canada to get to Pacific waters. Next was Frank, another Dutchman, a nervous kid who was just about to shed his teens. He had come to stay with his

brother. I was the third member of the trio.

Frank's proud possession was a beaten up Ford of uncertain age, that got him into more trouble than it was worth. The heater did not work, one of the windshield wipers kept getting stuck, the rear bumper was pried loose by one of Frank's latest attempts to take down hydro posts at high speed, one fender wanted to live a life of its own, and the floor had several holes that helped the passengers obtain accurate records of the chill factors at various speeds.

One February morning we drove into Temagami just as people began filling the streets to partake in the festival. The first thing we discovered as Frank turned sharply into a side street to park the car was that the right door, that was supposed to be permanently jammed, became permanently un-jammed. This put Gerry into an awkward position. He was trying to keep his upper body from being left on the street while I hung onto the seat of his pants to keep him in the car. When the car skidded to a stop, a conversation flared up between Frank and Gerry that was carried on in Dutch except for the frequent use of a well-known English profanity.

67

By the time we joined the crowds that lined both sides of the contest area everything was forgotten. Immediately our attention was taken up by the dog teams that lined up for the twenty-four mile race through the bush. Whips cracked, dogs yelped, men shouted. Suddenly they were off and disappeared amongst the snowladen trees. Other people took their places to compete in the tea boiling contest. A group of women and men hurriedly skinned muskrats for the honor of becoming Northern Ontario's fastest skinner. Lumberjacks pitted their muscles against each other, trying to be first in a log cutting contest. Out on the frozen lake cars raced in circles. Men came running to the fenced off area on snowshoes with their shirts undone, bodies steaming and their beards crusted with ice. Most fascinating were the Indians who had come from as far away as the Six Nations Reserve to entertain everybody with their tribal dances.

With so much going on it was no wonder that it was late in the afternoon when we felt the first hunger pangs and decided to look for a camping place. When it began to snow we drove to a nearby Provincial campsite that was already buried under fourteen inches (36 cm) of snow. Frank manoeuvered the car past the gate in search of a camping place but before we reached one we all found out that worn summer tires do not grip in deep snow. Gerry mumbled something in Dutch and left to set up the tent and start a fire. Frank and I labored to get the car freed and turned around. When it was finally done I was famished and looked forward to a hefty meal. My heart sank when I looked at our supplies. They consisted of frozen eggs, frozen bread, frozen butter, frozen soup, frozen beans, and very stiff bacon. The meal became an ordeal for our lips and stomachs. While my teeth hurt from eating bread that was saturated with ice crystals my stomach winced every time I gulped down food that was too hot. On top of everything it turned dark quickly and we had to let the fire go out because we could not find more wood. There was nothing left to do but to crawl into our sleeping bags and call it an early night since nobody was in the mood for an

68

intelligent conversation.

In the morning I was wiser for at least two reasons. First, when camping in the winter never use a summer sleeping bag. Secondly, when sleeping in the winter in a summer sleeping bag do not take off your wet corduroy pants and leave them lying in the

tent overnight. It is absolute misery to try to put on a pair of pants whose leggings are frozen together. You should also leave your boots on overnight or else you may find it takes a lot of skill and patience to feed wires instead of laces through the eyelets.

Gerry, with his infinite survival skills, managed to get another fire going so at least we could warm up again. Breakfast was much the same as the supper the night before, except shorter. None of us wanted to see more of the Winter Carnival but to get home quickly to a warm room and a good meal. It was not until we took the tent down that we noticed another ten inches

of snow had fallen during the night. Then we remembered the car and expected trouble. For the longest time the motor ignored Franks's coaxing and cursing until it had pity on us and coughed itself into a purr. While the car warmed up we took turns in shovelling it free with a piece of plywood hoping that the powdered snow would not be too much a force of resistance for the old jalopy.

These were not the happiest times for observing wildlife but perhaps the right time for a wild animal to observe us.

While we had been busy with the car a snowy owl had taken up an observation position on a birch tree twenty feet away. It could have been there all morning. Even if we had been completely silent we would not have heard its wings beat the air. With its fluffy wing feathers it approaches silently and delivers a soundless death to any prey.

I trudged through the snow to get a close look at the bird. Since it cannot roll its eyes it had to bend its head to follow me until it was looking almost straight down at me.

Its heavily feathered feet clutched a branch; its sharp claws curving around it. While I had cold feet,

and my legs stuck in wet pants that made me shiver all over, the owl was obviously warm and cozy, dressed for winter much better than any man could ever be. The soft white feathers protected it against the cold air. The feathers covered everything except the talons, the short hooked beak and its big round eyes, and protected her from any perils of the arctic weather.

Home of the snowy owl is the far north, but it will go to the southern latitudes during extremely harsh winters when food supply is scarce. Its feathery coat gives the impression of a bulky bird. Even the eyelids are covered with tiny feathers and can be moved independently. They also act like blinds shutting out unwanted light. The snowy owl hunts by day unlike most other owls, which are nocturnal. This may be an adaptation to the arctic daylight that shines continuously for months during the summer. The snowy owl feeds on rodents, mostly lemmings living in the tundra, but is most skillful in catching birds as well. One of the most amazing things is the owl's ability to swallow small animals whole. Larger animals, such as hares, are ripped apart with the powerful beak. All indigestible matter is later coughed up in pellets, where upon examination one can tell what has been eaten.

The snowy owl's nest is found on the ground in the tundra. It is lined with moss and feathers and contains between 5 to 10 eggs. They are incubated by the female, who has heavier markings of dots and bars on her folded wings and breast than the male. Usually

71

it takes thirty days for the chicks to hatch. They are covered with thick white down for the first 10 days. Then an even thicker grey plumage grows. Only three or four of the chicks will survive, the others often being the victims of other predators, their nest mates, or possibly the climate.

In the later years I had many chances to observe other owls, but I have always found the snowy owl the most attractive of the many owls that live in Canada.

At one time I observed how an owl scoops up its prey. I was sitting in a clearing enjoying a rest during a snowshoe hike. It was a cloudy day and fairly warm. Perhaps that was the reason a field mouse dug through the snow and crawled up a knotted pine branch that stuck out from a marred trunk. The mouse sniffed the bark and balanced on the jagged top of the branch. Maybe it sensed the danger or found the lofty spot not spacious enough. The mouse started for the hole it had made in the snow, but it was too late. A large white blob suddenly entered my field of vision. It was a snowy owl that swooped silently on wings which

spread some five feet from tip to tip. A short distance away, when it was level with the mouse, the owl made a number of simultaneous body adjustments, reaching forward with its legs, and extending its needle sharp talons, ready to sink them into the mouse's soft flesh. The beak was pressed against its chest because the head was bent downwards to allow its eyes to stay on target. It turned the outside edge of its wings down to brake the speed of flight, which gave it more time for

the final grab and the bird's claws pierced the mouse. Then wings, body, tail, and head adjusted to the upward flight that lifted it above the trees where it disappeared. The only sound that I heard throughout was the final squeal of the unfortunate mouse.

Owls often visit North Bay and one came so regularly over a period of years that the townspeople lovingly named it Orville.

Another visitor from the North was a hawk owl that I saw one day sitting in an apple tree. Much smaller in size and colored like a hawk it seemed to be unafraid of humans. It let me come almost within reach before it flew away.

Few animals have been linked mystically with man as much as the owl. It is supposed to be able to visit the world of the ghosts and return to the world of the living with messages. Some people see in an owl the summary of all wisdom. My own grandmother believed that a person hearing an owl hoot would lose a friend or relative in death. If the person was foolish enough to answer the owl's call he, or she, would die soon after.

The Indians regarded the owl with deep reverence and believed that the spirits of all medicine men passed at death into owls. It was a good omen to see an owl before a hunt or before fishing for salmon. Some Indians tell a story that explains how the owl obtained its crooked beak.

In the days of old the owl used to have a straight beak much like the raven's. Then the owl was often daring and careless, spending much of its time in foolish chases and flights. Once it saw a hare and went after it with great speed, but the hare quickly changed direction with the owl just a beak-length behind it. The foolish bird was unable to do the same and crashed into a rock. Since that time the owl wears a crooked beak and has become less daring.

THE PORCUPINE

Much of the ground was still covered by
hard-packed snow when the ice on the lakes broke.
Below ground spring began to stir in plants and
grasses that pushed upwards and the sap in the trees
rose. It was April in Northern Ontario: maple syrup
time. This was the time I had waited for, ever since I
first tasted nature's gift dripping from a stack of
pancakes at breakfast. Now I was to see how the
syrup was made.

A few miles south of Lake Nipissing my landlady's
brother owned a forest of maples. There I watched for

the first time the sap drip into pails that hung on every one of the twelve hundred trees. A hole was bored into each tree and fitted with a small tap. A pail with a lid was hung under the tap. The lid had a hole that let the tree deposit the sap at its own speed, while keeping out dirt. Then it was up to the farmer to check when the pail was full. He came round with a team of horses that pulled a short heavy sleigh carrying a big container. As soon as it was full with the watery liquid the farmer took it to a boiler shed at the bottom of the hill. There the sap was boiled day and night until it reached its sticky form and brown color. Frequent tasting by visitors and family members, especially the younger ones, established the right time to ladle it into one gallon cans which were then distributed to customers and dealers.

The familiar team of horses and sleigh was a victim of progress as now miles of plastic hoses are attached to each tree's tap and connected with the mainline that is as thick as a garden hose. In this mainline the sap runs straight down to the boiler shed.

On the edge of the maple forest an older tree leaned towards the fields that sloped down to the farmyard. Discarded wheels and sections of wagons were piled against it and made an ideal set in the mild sunshine. I took a pail from the nearest tree that was not keen on giving up much of its liquid and sat down on it to enjoy the view. While the sap dripped at pulse rate I tasted it expecting it to be very sweet. There was only a faint sweetness and a trace of an unknown flavor in the watery substance. It was a lovely thirst quencher, but a long way from being syrup.

From the boiler hut the voices of two men could be heard while the steam that was blown here and there by a playful wind often obscured the shack. Two dogs yapped at each other during a chase through the mud near the barn, a woodpecker knocked, and in the dense bush behind the farm a ruffled grouse drummed an escalating rhythm with its wings that sounded like an old tractor being started. Occasionally something shuffled behind one of the trees and at one time when I expected to see a rabbit, a most peculiar animal

77

ambled into view. First, I saw four long claws on a hairy paw reaching around a tree one foot above the snow. There it remained for a moment than slid up and down as if searching for a particular spot. The claw disappeared and seconds later a bulky little mound of bristly hair waddled around the tree. With every step it took, the porcupine's hair and spines quivered as if they were loosely stuck into the skin. In reality this is

due to the fact that the muscles surrounding each quill are kept relaxed. Only if the animal is threatened will these muscles tighten and erect the quills, much to the displeasure of any animal that sees in the porcupine either food or a playmate.

The porcupine is a peaceful rodent that normally goes about its business in a manner that evokes pity in the viewer for the seemingly dull-witted animal.

This one meandered from tree to tree quite unconcerned by my presence, keeping its nose close to the ground. The stubby, short legs were placed far apart, probably to facilitate tree climbing. It kept its head arched and its tail off the ground all the time. It seemed to visit the trees for no particular reason. Perhaps the invisible signs of spring had inspired the porcupine to search for early shoots or something that represented a bit of a change from its winter diet of

78

the inner bark of pines and other coniferous trees.

After a while it sat down on its haunches and sniffed the air. A piece of plywood that stuck out from the heap of farm paraphenalia I was sitting on seemed to attract the rodent's attention. Perhaps it was the resin used to glue the layers of wood together, or perhaps it had been lying on a salty road before it was carried here, or maybe dogs had urinated on it. It is often hard to tell what takes a porcupine's fancy for gnawing on things. They have been reported to chew on aluminum plates, tin cans, hoses, saddles, cottages, toilet seats, steps, fences, tools, tires, boots, paddles, boats, and all kinds of things that have been handled by man. It is assumed that even the slightest salt content will attract a porcupine. Unfortunately this leads to a high death toll on highways during the winter.

While the porcupine strained to reach the piece of plywood, I had ample chance to notice its long hair that stood almost perpendicular to the body, giving the animal an appearance as if it had just been through an electric storm. A heavy underfur of short darkbrown hair served as insulation against cold. There were no quills on the belly nor on the leathery soles of the feet. On its back and sides they were hidden under the hair, except for some telltale ends. Halfway down the back to the tail the quills thickened in density and size. The tail aided the animal in keeping stable as it reached out for the plywood but it is also a formidable weapon that teaches inquisitive or hungry animals an impressive lesson.

The blunt end of its face showed two large nostrils, sensitive tools in the world of smells. The upper lip was drawn back and quivered slightly. The mouth was open and exposed the upper and lower incisor teeth. On the cheeks a number of fine whiskers could be seen, while the eyes were small and needed wiping. A mixture of short quills and wiry hair served as eyebrows and extended down into sideburns that covered each cheek. The little round ears were almost lost in the maze of stringy hair and quills that started on the ridge of the nose and ended on the back.

The porcupine stretched and stretched but the plywood remained out of reach. Even extending the four claws of a front paw failed to bring the board down. The animal sank back to the ground and just stood there as if lost in thought. Then it glanced at me as if to say 'here is some more for you to look at', turned around, arched its back and deposited some droppings for my examination. All its movements were slow, yet deliberate.

No matter how much one minds his own business and tries to live a peaceful existence, someone is bound to come along and disapprove of it.

I was the first villain to disrupt its peaceful mood. To find out how sensitive the quills were, I took a stick and gently touched its back. Immediately the quills were erected while the tail swatted upwards. It stood there for a moment, head bowed, stiff-legged, tense, and ready for another swipe. My curiosity satisfied, I waited for the animal to relax, but Tippy, the little dog that suddenly appeared, had other plans. He was a smart dog who often sought his master's special affection by presenting him with rats, snakes, squirrels, and even the occasional skunk. Tippy was outfitted with wits that were ideal for a life in a cozy home as well as in the wilderness. I am sure he knew that a porcupine does not make a good gift, but did present a terrific chance to test reflexes. Tippy tried nipping the porcupine's face but the rodent quickly whirled around and lashed out at him with its spiky club. Tippy darted away quickly enough to avoid a devastating blow. The dog barked and snarled, but no matter how often he circled to face it, he always ended up confronting the rear end. For a while they kept each other at bay, then the porcupine decided it had

had enough. It waddled to a tree while keeping the quills and tail at ready. Tippy merrily watched as it got up on its haunches and embraced the trunk. It sunk its claws into the bark and drew its hind legs under its body. Then it gripped the tree higher up and pulled itself up with vigorous movements, displaying skill and strength in climbing that seemed foreign to its otherwise sluggish nature. The feet and the tail proved to be ideal tools for climbing. It had four claws on each front paw and five on each hind foot, which gave it a good hold in the bark. The barbs on the underside of the tail acted like spikes against the bark and supported the body as it climbed higher and higher. When it reached a strong branch it settled in

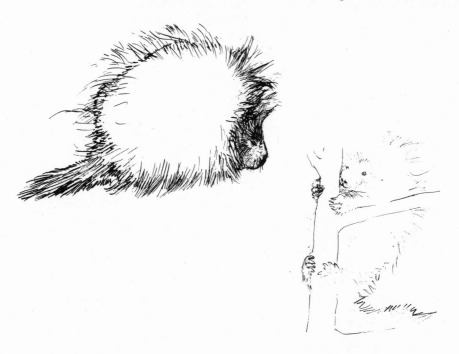

the fork for a rest. In the trees it gets its food and can live an almost undisturbed life. It has many enemies, but few succeed in harming it. Weasels, wolverines, bears, cougars, and even wolves have been found with quills in their body, mostly in the face. One such unfortunate wolf caused some commotion in an Indian neighbourhood one winter afternoon. It was observed through a kitchen window as it left the cover of the bush and leapt through the deep snow heading for the back of the house. This is such unusual behavior for a wolf that one of the men fetched a rifle and shot the animal only twenty feet from the backdoor. The wolf had apparently attempted to kill a porcupine and some of the quills had worked themselves into the gum through the cheeks, making it impossible to hunt or to swallow. The animal was crazed with starvation and probably attracted by the odor of the garbage, had desperately tried to get some food.

The fisher is the most adept porcupine hunter and much of its diet consists of porcupine meat. It is quick and usually manages to get at the rodent's soft and unprotected belly.

A pair of coyotes were once observed by a trapper as they attacked a porcupine. While one of them confused the porcupine by circling it and eventually caused it to leap around, the other sneaked up and flipped it on its back, rendering it defenseless.

The Indians have an explanation for the porcupine's habit of resting in tree tops. As in many of their legends this one also happened at a time when all animals spoke the same language. Then the porcupine was very lazy and spent most of its time sleeping, while the beaver was always collecting food for his storage den. One time Porcupine ate all of the beaver's supplies instead of looking for its own. The beaver was so enraged that he attempted to beat up the porcupine, who quickly rolled into a spiky ball that pierced beaver's tender skin. Beaver however got his revenge next time Porcupine was sleeping and carried him off to a rocky island. The porcupine would have starved to death, but the water froze and he was able to return and in turn overpowered the beaver and carried him to the highest tree. Unable to climb down by himself the beaver could think of only one thing to do. Slowly and carefully he gnawed for days to shorten the tree until after each labor there was nothing left of it but a pile of chips. To this day the porcupine and the beaver do not speak to each other. Often one may see the porcupine high in a tree longingly looking at the beaver's food supply, though no longer thinking of stealing some. Meanwhile the beaver continues to cut down trees wherever he can, so that he will never be taken to the top by the porcupine again.

83

BLACKFLIES

There comes the time for every immigrant to exchange old familiar views for new ones. The speed with which one adapts to 'When in Rome, do as the Romans!' depends on a person's open-mindedness and how much or how little he mixes with people of his own cultural background. I was convinced that Canada's life-style would have little influence on my principles, one of which was that no one should buy more than he earned.

During the second summer of my new life in Canada a used car dealer helped me to see this idea in a different light. He reasoned that a person who has a steady income and no other financial commitments is depriving himself of hours of pleasure and comfort by saving first and buying last. The principle remained intact if one obtained the goods first and saved later, he insisted. Somehow all this seemed logical and reasonable as I sat behind the steering wheel of a car on the lot and imagined myself driving through Canada. The places I could go, and things I could see seemed nearer already. He offered to put a roof rack on without charge and tipped the scales of my indecisiveness with the words: "With a car and a canoe the wide open spaces of this land are yours."

And so they were.

Within a few weeks I made numerous trips that showed me the diversity of the country. From the awesome moonscape of Sudbury to a secluded little shack of a grouchy old loner whose sign read: 'Trespassers will be shot at!' the land shrank and it became more familiar and more and more fascinating.

Often I drove along the highway enthralled by the vastness of the country. Occasionally I stopped, shut off the engine and listened to the sounds of the wilderness that lined the road on both sides. Sometimes I followed a bush road to see what lay waiting at the other end and discovered the great sadness that lived in abandoned shacks and farm buildings. Though it seemed aimless driving, it was not without purpose. I wanted to find the place where the bear grumbled, the wolves howled, the deer grazed, the wind chased the fog, and the trees spoke with the rocks. I wanted to find the place and that hour of magic that would speak to me and tell me that I had not come to this country to be lost in it, but to belong.

Once I followed a bush road until it split into two. The section that showed signs of recent use led me to a small lake. Its rippled dark surface had tempted many fishermen who had left an assortment of rubbish in a small clearing. I drove back to the road fork and parked my car. Patches of young growth covered the second road but since it was not too dense to walk through, I hiked down until I reached an old mine. Apparently it had been out of order for years. Its dark ominous entrance was supported by hand-cut logs and boards. Rafters lay around and a few feet inside, the water on the ground mingled with the blackness beyond. It was cool there and for a while I stood and listened to the drops tick off time in a forgotten world.

A ray of sunshine painted a golden crown on a frog that sat in the water by my feet and memories of long lost childhood days swayed before my eyes. As a child I believed in fairy tales and turning around I was

stunned by what I saw. For a few moments I thought that one of my most ardent childhood wishes had come true. Before me in the brilliant sunlight stretched a road of silver. Though I wished this fairy tale were true, I knew, instead, that there had to be a logical explanation. I walked up the road a bit and looked back to the mine and the silver disappeared. Facing the sun, the road sparkled and glittered again, but as soon as I turned towards the mine the road was bare and drab. When I ran my fingers over the ground I discovered the real reason. The road was covered with specks of mica that glittered like silver against the sunlight. I scrambled up the short slope into which the mine's entrance was dug and sat down to look at the spectacle.

To my left a trio of birches leaned towards each other and swayed like drunkards. A lizard scurried into a shaded rock crevice. A grouse stepped onto the road, strutted across and disappeared into the shrubs. I dug up a handful of earth and looked at it. It was black and rich, nourished by leaves and rotting wood that had fallen to the ground over thousands of years. This was the soil that existed long before man ever set foot on it.

The gift of this land was not the wisdom of Socrates, nor the clever sayings of Laotse. Here, wisdom was of a different kind. It was silent, harsher and greater. This wisdom was not being discussed by learned philosophers and nowhere was it to be found in books because only here was it alive. It lived in every grain of soil, in every pine needle, in every insect, in every cloud. The wind carried it from the Arctic and brought it as a message to every man who cared to listen: 'We have been here long before you: we will be here long after you have gone!' This was life without time.

From a nearby swamp a cloud of blackflies arose thirsting for the blood of mammals. They were completely disrespectful of my contemplative mood. Soundlessly without warning and they fell upon me. At first I thought I was able to ward them off, but I soon realized the futility of my intentions as they sought to draw blood from my arms, legs, cheeks, and neck. Then they attacked me in such great numbers that I became angry. They literally settled in my nose, my eyes and ears and I got little relief from a branch that I vigorously thrust about in desperate defence. I raced to the car hoping to leave them behind. Instead they followed me and induced moments of panic as I fumbled to unlock the door and start the car. Having little respect for the incompatibility of high speed and backroads I had only one wish: to leave those tiny monsters behind. Finally I reached the highway and with the windows wide open the sixty miles per hour draft persuaded the most persistent ones to part. After a while I calmed down, pulled the car over into the side of the road and with the windows tightly shut contemplated the experience. I had heard of whole herds of caribou breaking out in stampedes because of blackflies. Moose are known to swim far into lakes to escape them and domestic animals have crashed through fences crazed by the stings of these insects. I have seen dogs blindly seek refuge in rivers and even dig holes in the ground, all in an effort to escape the pesty insects.

As if nothing had happened, a blackfly suddenly emerged from behind a sun visor and buzzed down the windshield. I caught it and noticed it was no bigger than an eighth of an inch (3mm), as I studied it under the magnifying glass. The humpbacked miniature monster struggled to free itself, but when I brought my finger near her head the wings stopped beating and the two antenna feelers searched the skin to locate a vein. Even though the female needs blood to perpetuate her species, I found it hard to generate any sympathy for blackfly motherhood at that time. Blood is essential for the development of her eggs, that are deposited in the water, often on sticks, rocks, and floating debris.

To get at the blood she has razor sharp scissors on her mandibles and with these tools even the tough skin of an old bull moose can be easily pierced.

She selected a spot on the soft part on one of my fingers and in the interest of discovery I successfully

88

fought back the urge to squash her as soon as I felt she
was cutting my skin. Although I could not really see
how she deposited saliva into the wound, I knew that
she was doing it. This is the fluid that contains
chemicals which cause swelling, itching and the
burning that lasts for days, even weeks. As soon as the
incision is made the blackfly also injects a liquid that
acts as a narcotic and anti blood clotting agent. This is
important as it prevents her feeding tube from getting
clogged. With all these chemicals that come into the
body from only one fly it is easy to understand why
one can experience nausea, headaches, swollen glands,
and other discomforts if bitten by many blackflies.
Sometimes medical treatment or hospitalization is
necessary. The death of horses and cows had been

reported three hours after a sudden blackfly attack.

As many as sixty thousand eggs can be laid on one leaf by a swarm of blackflies. Scientists who observed a short stretch of river estimated there were some seven billion blackfly larvae in the area. It takes a larva two to ten weeks to reach maturity, depending on water temperatures. By a special hooklike device the larva can withstand strong currents and feed on the plankton and other microscopic food that drifts in

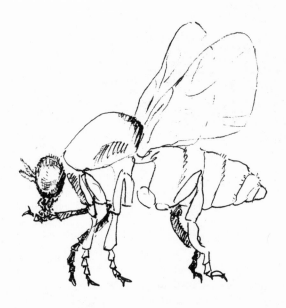

the water. Shortly before its escape time the new fly breaks from its tiny underwater prison with the help of an air bubble that raises it to the surface. The speed of the rising bubble helps the blackfly to take off into the air. Although a great number of larvae and mature flies are destroyed by fish, birds and other insects there are still enough left to make every outdoor trip a potential introduction to purgatory. For years pesticides, notably DDT, were used to force some kind of control on them, but the chemicals proved to be harmful to the ecosystem. Perhaps an oldtimer's remedy is still the best protection against the flies. He insists that his concoction is a sure way of keeping the flies away and I suspect everybody else as well.

Take three parts of pine tar, two parts of castor oil, one part of linseed oil, bring to a light boil and simmer over a small flame for forty-five minutes. When cool,rub it liberally into the parts of your body that are most vulnerable to attacks and never wash it off. If the glaze weakens repeat the application and in time you will be able to walk through any swamp without even getting to look at a blackfly.

Some of the Native Indians insist that blackflies do not like foul language and every white outdoorsman should at least know two or three words in Cree or Ojibwa.

The storytellers of the old legends used to explain the birth of blackflies this way:

When time was very young there lived a giant spider whose eyes emitted deadly rays, from his mouth came poisonous gases and his countless legs were fitted with slashing claws. The monster thrived on human flesh and went about ravaging the earth and decimating mankind. The bravest warriors failed in slaying it and in the end only a few people survived. They prayed to the Great Spirit for help and he dispatched a supernatural bird to rid mankind of the monster. The bird man built a huge pile of dry wood and tinder and hid in it. From there he teased the short-tempered spider monster which soon found itself stuck in the wood pile trying to get at the man. The bird man then set fire to the logs and branches and quickly crawled away. In no time the flames reached to the sky and roared for days until only a huge pile of ashes lay smouldering on the ground. Great must have been the monster's power, for as the ashes cooled it appeared again in the form of tiny creatures with wings. Silently they rose by the thousands, each carrying sharp claws in front of their mouths. Though the power to kill people was taken from the monster, its descendents were able to torment them by sucking their blood and depositing a tiny speck of ash in the small wound that would itch for a long time.

91

THE VARYING HARE

When the hills are ablaze with autumn colors, the days still bright and warm, the shadows cool and the bush is without stinging insects to bother either beast or human, then millions of tiny spiders spin silvery threads and hang them up everywhere. The spiders love this time of the year and perhaps are trying to tie the days of the Indian summer to the land with their threads so they may stay a little longer. A person who has been in the Canadian North without experiencing an Indian summer knows only half a wilderness. Those days humble a man and make him aware of God's love for nature. It seems to ooze out from every rock and

92

plant, every lump of earth, every drop of water. It seems to quiver in the grass, sway in the trees and sing with the birds. It seems to surge through one's veins and one realizes that each day spent without love is a day of loss.

The lure of the brilliantly colored autumn has inspired poets, writers, composers, painters, and nature lovers. Perhaps it also inspired a romantically inclined native story teller of long ago to think up the story of why the leaves of trees change colors in the autumn.

The legend goes that in ancient times a huge rainbow arched the Northern hemisphere after a severe storm. Lighting Snake carelessly dispatched a light bolt that broke the rainbow into millions of pieces and scattered them over trees and shrubs. The Great Spirit saw what happened and, delighted by the new beauty of the land, decided that from that day on all green leaves should take on the colors of the rainbow every autumn before they fall to the ground.

One evening I watched the golden hour of sunset while I sat on a rock that could well have been the very same place where the inspired Great Spirit beautified the world. I over-looked a little lake that in part sparkled with the diamond-studded minutes of the last hour of the day, and in part was already hidden by the far reaching shadows of the approaching night. The dark figure of a man sitting in a boat came slithering through the water, quietly driven by a double-bladed paddle. The man beached his slim kayak nearby and turned to watch the sun compete with the flaming bush. Finally, when the brilliance of the sunset was pushed over the horizon by the coming night he faced me and muttered: "What a beautiful country."

The brief meeting initiated a long friendship between the Dutch immigrant Gerry Beerkens and myself. It was he who made the used car salesman's prophecy come true by selling me one of the kayaks that he built himself.

Now I had a car, camping gear, a boat and in Gerry a companion who in the following years shared my wilderness experiences. Gerry not only loved the bush emotionally but also in a very practical way. Hunting and fishing were useful things to him. As long as one did not abuse nature and aligned one's common sense with religion one could do no wrong. It was a simple, honest and healthy philosophy. In hunting terms it means that you do not shoot a squirrel because you want to try out your gun; when you go rabbit hunting you do not aim a twenty-two calibre rifle at a deer. You must know the animals, the bush, and your weapon. Above all, you must remember, once you pull the trigger you cannot stop the bullet. Do not spoil, do not waste, do not litter, go silently into nature and she will reward you with a good time.

Before I could get proficient in handling a kayak, the Indian summer was over and cold winds and low temperatures made boating unsafe. Gerry thought it best to postpone the training until next spring and in the meantime we should plan a trip. Would I like to

have dinner with him and his family on Sunday? We could discuss the trip afterwords. As a bachelor whose cooking skills consisted of warming TV dinners or buying fried chicken I was always keen on highlighting my food intake with homemade meals and quickly accepted. How did I like rabbit stew? Well, it sounded great to me, though I could not recall having tasted it before. Would I help him prepare the meal? How could I refuse? I wondered what I would help him with. That was easy, he conceded, we were to meet Saturday morning and first go rabbit hunting.

On Saturday we found ourselves after a two-hour

94

drive somewhere in the bush off Highway 63 looking
for hares. I carried a borrowed .22 calibre rifle and this
was the first time in my life that I was out hunting
animals for food. An earlier snowfall had left white
patches in the drab pre-winter shades of brown and
only the fir and pine trees and the sky were colored in
greens and blues. We walked about 40 feet apart, still
in view of each other. Occasionally the undergrowth
was too heavy, therefore we whistled to each other so
we would not lose one another. Under some
poor-looking birches lay batches of paperlike bark and
I would have walked past had I not seen something

move. It was a snowshoe hare in his white winter dress that was only showing a bit of black along the ears. Slowly I walked towards it and though I was sure it saw me, it did not move. I was so close that it seemed unfair to shoot. I knew it did not have a chance, yet I raised the gun.

A childhood memory rose before my eyes and I saw my father bringing two dead rabbits home from a hunt. I must have been only four years old, as shortly after my father left never to return. He was trying to explain something to me - that the rabbits were dead - but I did not understand. Later I stole into the pantry and watched the motionless bodies of the animals as they lay stretched out on the shelf. I asked them to move, breathe, sit up, wiggle their noses, but they did not seem to hear. I stroked them and blew my breath into their soft fur and finally picked one up and ordered it to stand. My hand felt wet and when I looked at it I saw blood. I realized when blood runs out of a being, life runs out as well and sensed the meaning of death. The memory vanished and I found myself debating whether to pull the trigger or not. I could see the hare's sides rise and fall with every breath it took. Even its nose had grown a fine layer of white hair and its long whiskers were embedded in mottled cheeks. The only other patches of color were above its eyes that seemed half closed. There was a bit of a sandy hue in the soft white fur which was caused by the brown undercoat shining through. The legs were tucked under its bunched-up body exposing only the outer tips of the toes that were covered with long yellowish hair. The hare had been sleeping as I found out the moment I stepped on a twig. With the crackle it sprang into action. Though it was gone quickly I noticed the tiny tail because its tip was covered with black hair.

I walked on wishing the hare a long life, but no hare or rabbit is blessed with old age. Most fur bearing mammals as well as larger hawks, owls, and the eagles prey upon the family of hares and rabbits. To native people they are an important part of their diet. They are also susceptible to many devastating diseases and therefore it is no wonder that a hare will seldom see his second birthday come around. Perhaps for this reason hares feel compelled to breed three or four times a year producing a litter of four young on an average each time. After a gestation period of thirty-eight days the young are born fully furred and with their eyes open. They begin to feed on tender bits of greens almost right away. At intervals of ten years the hares display behaviour that has puzzled scientists for several years. They reach tremendous abundance, which is followed by a sharp decline in population that may last for two or three years. Predatory animals seem to have little effect on the mysterious cycle of population explosion and decline.

Although the hare can travel at a speed of more than twenty-six miles per hour it is often able to escape predators by remaining still. When it has to run from them in the winter, its long furry paws give him sure footing as they act like snowshoes on top of deep, soft snow. Because of these paws it is also called the snowshoe hare. Usually a slight hollow in the ground in dense bush or under low hanging branches serves it as a comfortable place to spend the day. When evening arrives, hunger drives it to seek out grasses, herbs, leaves or the bark of trees.

The name varying hare is derived from its ability to change colors with the seasons. In the summer its reddish brown fur blends in with the colors of the earth hiding it as effectively from its enemies as its winter coat of dense white hair does, after the snow has arrived. The Indians say it is a reward for a good deed which the hare once performed for a supernatural being.

It was during a long drought when the animals had disappeared and had left the Indian people with the threat of starvation. Even the best hunters returned to their camps empty handed. A supernatural being, a benefactor of Indian people, heard of their plight and came to help them. He offered to find out where all the animals had gone and to lead the people to them. After weeks of travelling he came to a great river and was sure that this was the place where the animals were now hiding. During the night snow fell and next day he saw tracks all over it. He set out to return to his people, but became snow-blind. Completely lost, he stumbled upon a brown hare that guided him back to the people, whose lives were saved by the kindness of the little animal. As a reward the supernatural being gave the hare the gift to change its brown coat into a white coat with the first snowfall.

During our hunt I spotted five more hares and each time called on Gerry to do what we both had come for. I just could not bring myself to shoot them. Gerry's patience and understanding ran out when I yelled at him for the sixth time: "Here is another one." "Why did you come hunting?" he hollered back, "Shoot it or stay home." It was obvious that he though I was better off staying in the city and to keep on dreaming of living in the wilderness than to subject myself to the toughness that one has to cope with in wilderness living. I knew he was right and aimed my gun at the easy target on the snowless ground uttering a silent apology to the hare. The rifle coughed harshly and the hare jumped into the air, fell back with quivering limbs, then relaxed. The sight of its lifeless body saddened me, yet on Sunday evening when Gerry's wife served the delicious stew I behaved as if happiness was a matter of digestion. Like the Indians who lived a grateful life I felt compelled to thank the hares silently for being my food.

I realized that life in the wilderness required a different moral attitude.

SOME BIRDS

During the winter, Gerry and I spent the occasional evening planning our first kayak trip and decided to enter the Jocko River at Highway 63. Twelve miles downstream the Jocko joined the Ottawa river, where we would camp overnight. Then we would continue on to the hydro dam just outside Mattawa, where a friend would pick us up by car to take us back to North Bay.

Spring comes quickly in Northern Ontario and though I was eagerly awaiting a period of warmer days, I felt I had been away for some time when the new leaves appeared almost overnight. Arrangements were made with our friend to drop us off at the Jocko River and again to pick us up two days later at the power dam. Before long we were paddling away from the bridge just as daylight flooded the sky. Inside our kayaks were provisions for our new adventure, including fishing rods, the necessary camping utensils,

100

rain ponchos, and a large plastic sheet instead of my heavy canvas tent.

The river was like a tame little stream that wound itself through the dense bush, the water was cold and it looked unfriendly to me so early in the morning. I still felt unsure about handling fast flowing stretches, but my anxiety was soon wiped away by my effort to keep up with Gerry. With powerful strokes he kept his kayak speeding along, displaying strength and skill that I tried to copy. The trees and shrubs lined the shore like a curious crowd and soon it seemed we had entered a land untouched by human hand. Gradually the water became shallower and boulders began to show through the surface. Then, through the silent wilderness, the roar of water cascading over rocks kept getting louder and louder. The faster the water flowed the less room the rocks left for our kayaks. Then we had to face the unavoidable: a portage. For the next two hours we labored and wrestled with the kayaks, bumping and bruising ourselves as we slid off wet rocks, or fell into deep cracks. Since the bush was so dense that it was impossible to find an opening, we had no other choice but to follow the river. My shoulders ached and my legs hurt. I was wet up to the waist and just when I really needed a break from this obstacle course, the shallow river squeezed between huge boulders. We were lucky we had come at a time when the water level was low or else we would have had to spend hours cutting a trail through the bush to bypass the rapids. This way we could at least stay with the river even though it meant lowering each kayak with ropes. Finally, the rocks disappeared and

101

ahead the water was once again navigable. We paddled on until the river widened and the trees stood back to let the sunshine flood a gravel bank. There we rested. We both stretched out on the ground and while I stared into the sky I became aware of the voices of songbirds.

From the branches of an old weather-scarred fir tree came the call of a white-throated sparrow. Its song quickly warmed my heart and I knew then that we were being rewarded for our bruises and our fatigue. Gerry, too, came back to life and brewed some tea, while I tried to spot the bird with binoculars.

The sparrows liquid tune consisted of four notes which it repeated again and again like a flute player in practice. The world was alive this morning and the more I listened, the more birds I heard. No doubt there were many travellers on their way north. A group of warblers flittered through the bush like wandering minstrels, thrushes called each other, woodpeckers hammered, a squirrel chattered, somewhere overhead ducks quacked a message, and like an indignant neighbor a raven complained with harsh croaking. Finally, I spotted the sparrow as it left its lofty perch and swooped down into a shrub on the opposite shore. It disappeared into the foliage and seconds later was back again with an obliging performance. It whistled its brief song repeatedly and put a lot of sweetness into the tune to make up for its lack of brilliance.

The members of the sparrow family resemble each other closely in general coloring, except for their individual markings.

The white-throated sparrow was given its name because of the white spot just below the beak. Two black stripes run from the beak across the head, separated by a dash of white. In front of each eye are spots of yellow that change to white above its eyes. Gray cheeks, a gray chest and a grey tummy that is dotted with rows of dark feathers make up its frontal attire. The wings are brown, streaked with darkbrown and show two white bars.

This lovely songbird takes up winter residence in

Mexico, though it has occasionally been seen in some parts of Canada during the winter.

After much searching I found its nest in the thickest part of the brush. It was lined with twigs and ferns and contained four pale blue eggs.

Refreshed with tea and sandwiches we continued our journey and found the bush retreating to give way to a marsh. There was no current anywhere but the water was deep and often branched out in side channels which we explored for signs of wildlife. We saw ducks, whose morning gossip was suddenly interrupted by our silent approach, muskrats that took to the black depth of water, perhaps feeling threatened by the paddles, and a beaver that turned into a narrow ditch without slapping its tail. Even a little snake took to the reeds that to us were off limits.

Nearby, a cattail stalk swayed under the weight of a red-winged blackbird that immediately after landing raised its wings to show off its brilliant scarlet patches on the shoulders and announced that this was its domain. Its cheerful and far carrying flute-like sounds were heard thoughout the marshes of southern Canada. This little fellow winters in the south as well, but arrives quite early to set up housekeeping—often before the ice melts from the lakes.

The marsh sounded like an echo chamber with

their calls. After some difficulty we found one of the nests. Five bluish-green eggs lay in it, spotted with purple and brown dots. As soon as we backed off, the female came shooting back, rearranged the eggs and sat down to watch us nervously. She contrasted with the shiny black of the male and was heavily streaked with browns, greys and reds.

We followed the main channel that broadened steadily until, stretched out ahead of us glistened the water of a small lake. For some time Gerry tried his luck fishing, but the fish were not biting. We trained the bows of our kayaks on a small opening in the bush where the river continued and paddled on. It was early afternoon and a lunch break was due. In about two hours we expected to reach the halfway point of our trip, where we planned to make camp. While we ladled out the soup, Gerry decided that we were in no rush and that we should both leisurely fish in the pools and bends as we drifted downstream.

Slowly we drifted with the river until again we could hear the sound of water falling over rocks. When we came to the rapids we left the boats and investigated them on foot. The water tumbled over a polished rock into a large pool and from there cascaded into quieter waters twenty feet below. Then the wide, slow Jocko river joined the broad Ottawa River. Gerry, keen as ever to have speckled trout for dinner, patiently tried out the various baits while I made camp. Except for the kayaks, which we later lowered over the rocks together, I fetched our gear, made a fire, cut some poles and draped the plastic sheet over them. I was quite content to look at the scenery and listen to the roar of the water. Nothing is more disheartening than to believe you are in a wilderness, untouched by man, and find beer bottles or tin cans. I dug a hole and buried the signs of earlier human visits. Then I played out the fantasies of a little boy who saw himself as an early-day pioneer. Gerry added a touch of reality to my imaginary time change as he appeared with four trout that fitted snugly into the frying pan. This free wilderness meal, in itself a new experience, brought an incident to mind that happened in Austria, where all land is privately owned.

As a teenager I went on a hike into the Alps with a friend. We crossed a meadow that belonged to the village pastor, together with the cows, the adjacent pine forest and the small mountain stream that gurgled down into the valley. We saw several rainbow trout floating motionless in the bubbly rock. My friend quickly fashioned a snare and lowered it around one of the fish, while I nervously checked that nobody witnessed this theft of holy property. He deprived the fish of its natural element with a sudden thrust of his arms, and there in the clover lay a beautiful trout flapping its tail. We retreated into the depth of the forest with our treasure and roasted it over a small fire according to the instructions of an adventure book that I had read earlier.

This time circumstances were different. Though the saying goes 'Stolen cherries are sweeter' I cannot say that stolen trout is tastier. There is a much greater quality of happiness to be experienced in sitting on the ground that belongs to no one and to watch the fish that one has just caught frying in butter and know nothing has to be done in secrecy.

When the heart is free, it recognizes the many subtle forms of happiness that offer lasting impressions.

The odors of the wilderness can be savored in their full strength, the sane beauty of nature can be marvelled at aloud and the inner self can fuse unhampered with grass, trees, water, sunlight and air.

After darkness had come we sat for hours, lost in pleasant conversation that was born by leisure, night and campfire. Like the smoke it drifted here and there.

Later, in our sleeping bags, we looked through our transparent tent and watched the stars and the darkness that reached deep into the universe until sleep blocked out everything.

My ears were unaccustomed to the pleasant melodies of birds that announce each day in the wilderness long before light comes and so I awakened next morning by their soft twitter that grew in strength and variety with every new sun ray.

Much of the beauty of nature lies in the sounds we hear and without the songs of birds the wilderness would not have the serenity and peacefulness that calms many a man's sorrow and diminishes his regrets.

One can easily understand the feelings of the Great Spirit as he once walked the forests long before the birds were created. An Indian legend tells us that he was sad, though he thought the world was really beautiful. One time during the autumn while he sat on a rock and watched deeply colored leaves sail through the air he became aware of the quietness in the forest. He thought how much more beautiful the world would be if these leaves would travel through the air all year long, so he gathered some of them. He changed them into small birds and ever since birds with feathers of all colors forever warm the hearts of men who travel the forests.

The hushed tune of a Swainson thrush caught my attention, just when I decided to get up and start a fire. It was very close by and so I remained in the cozy sleeping bag to listen as it practiced its part in a subdued manner like a musician who is about to perform on stage. Moments later it landed on the tent pole above our heads and delivered its well polished melody, that climbed higher and higher with every note.

Possibly delighted with our total attention it went over the same tune with new relish and then hopped

along the pole into the sunlight that, like the spotlights in a concert hall, illuminated the star in its full glory.

This small bird has the reputation of being very secretive. Its head and wings were olive in color, there were spots on the pearly white chest and a fluff of white covered its tummy. Around the eyes were rings of buff feathers and the same shade colored its cheeks. There are only slight differences in color tints and hues between subspecies that are found in both east and west Canada. Both sexes are colored alike. Sometimes during migration the Swainson thrush will seek the company of the hermit thrushes.

108

The nest of the Swainson thrush can be found in trees and shrubs a few feet off the ground and in it four or five pale blue eggs with brownish flecks are hatched in two weeks. Fourteen days afterwards the young fly away and it is quite possible to find the nest occupied by the parents raising a second brood before they embark on their journey to the south again.

The graceful visitor disappeared from the tent pole as quickly as it had come and then for the rest of the day lived up to its reputation of elusiveness,

Gerry fished most of the day, while I roamed the abandoned farm that we had discovered just beyond the ridge where the rapids roared. According to our calculations Mattawa was only fourteen miles down river. We had arranged with a friend to pick us up at six p.m. and all too soon the time came for us to lower the kayaks into the water again and start on the last part of the weekend trip, which we thought was only one hour of easy paddling.

The Ottawa River was a broad body of water that had no noticeable current and often gave the impression of an elongated lake. Hills on both sides undulated like giant tree-covered waves and hid the sun in the late afternoon and evening.

First we paddled without hurry knowing that soon we would be in the car that again would deliver us to civilization. After considerable time I noticed that Gerry's stroke became more powerful and quicker. I carried no watch, but some red clouds, just above the tree tops behind us showed the time was near sunset, which was well past six p.m. I called to Gerry to slow down and turn to shore where I could hear the bubbling of a creek. We both rested a little and refreshed with a cool drink from the brook, we started anew, hoping that behind the next bend we would see the hydro dam. But when we had reached the bend we saw the river disappear around a distant hill. By now it was obvious that we had made a mistake in calculating the distance between the delta of the Jocko and the power dam near Mattawa. I hoped that our friend with the car would have the patience to wait for us.

109

The light grew dimmer and the hills were like giant black monsters with spiked backs. For a little while we paddled in the centre of the river where the last daylight gathered, but soon it was pitch black. I was unable to orientate myself and often stopped paddling to listen for the swoosh of Gerry's paddle. Sometimes I called his name and found from his answer I had gone off course and when I asked him once how he knew in which direction to go he only replied: "I don't." This was obviously a case of the blind leading the blind. Hunger began to gnaw at my stomach and I was cold, tired and scared. We had often seen tree trunks in the water and I worried about hitting one. I called Gerry to find out if I was still going in the right direction, but there was no answer. I remained still and strained my ears after each call, but the silence was as deep as the night was black. I decided to paddle on until I either reached the dam or hit the shore. From somewhere the quick flash of a light darted by. Again it went by and since there was no road in the hills or on the shore it could only be someone's flashlight. I headed for it, paddling furiously, and called "Hello, hello". I was heard and guided to shore with the light. I found to my great relief that it was Gerry. He helped me ashore. I put on a heavy sweater and dug out my flashlight. Further search produced a small piece of bread and a can of sardines. Gerry insisted on going on while I was ready to simply drop to the ground right there. His watch showed it was midnight as we gulped down our meager late 'dinner' and once again squeezed into the kayaks. I paddled into the darkness like a robot and somehow managed to stay with Gerry. He was first to notice a light grow behind a hill. It spurred us on and though it grew stronger for some time it did not seem to get any closer. Worst of all, I began to feel sick. Battling with fatigue and the sardines in my stomach I finally made the last few yards to shore.

At five o'clock in the morning I collapsed into my bed at home and swore to myself I would never again go on another kayak trip.

It was an oath that I would soon forget.

110

THE MOOSE

During my third year in Canada I became aware of the tremendous size of a moose in a sad way. Driving to a friend's farm early one September morning I came upon the scene of a highway accident a few miles south of Callander. The night before, a thousand-pound (450kg.) bull moose had appeared in

111

his six-foot-high (2 meter-high) bulk so suddenly in front of a Provincial Police cruiser that the vehicle smashed into the big animal. The driver was instantly killed while his partner escaped with injuries.

The bull lay in the ditch, his huge carcass bearing mute evidence of being the largest antler-bearing member of the deer family. One of his legs seemed broken, it was hard to tell since they were covered by his body. The hooves were cloven and each must have been at least eight inches long (20 cm). His dark brown compact head rested on bunched up grass, the thick tongue had slithered out of the slightly open mouth. A mahagony-colored rack of antlers extended to a width of six feet. Each of the arm-thick main stems flattened into a palm that was fringed with points.

Only two weeks later the North Bay newspaper reported the death of fourteen moose in the White River area alone—all highway accidents. Winter had scarcely begun.

The moose, once also a resident of Western Europe, lives in the vast timberland that stretches from Alaska to Newfoundland. They are at home in mountain ranges, in the Northern tundra and live throughout the thousands of miles of lakes, muskegs, streams, and swamps of the boreal forest that stretches from coast to coast. Authorities claim there is a moose population of ½ million in Canada alone. The moose was an animal I always wanted to observe in the wild. My first opportunity came during another kayak trip with Gerry, shortly after I saw the dead bull beside the highway.

Again we started our trip from the bridge on Highway 11, but this time we paddled upstream to explore the upper Jocko river. The weekend coincided with those few days of the fall when the colors of the leaves reach a peak in brilliance. Usually frosty nights preceded this special time and so we were well prepared for cold nights, but since during the day it was quite hot we were soon shedding layer upon layer of clothing. Though the summer had dried out much of the river there was enough water left to form a narrow channel which led us without any problems to

112

a swamp a few miles upstream from our starting point. The main channel split up into several side arms and we followed one that seemed to have the swiftest current. It led us to a beaver lake. We approached the beaver dam and found someone had opened it up near the centre where the water spilled into a pool four feet below. The opening seemed deep enough to Gerry to paddle the kayak through. It was, but when the bow of his kayak hit the surface of the pool, the stern refused to slide on and Gerry was caught in midair. Unable to reach bottom with his paddle to push himself free, he decided to use his upper body, and did what a rider would do to a stubborn mule. "Come on!" he yelled while he jerked his heavy frame forward. The kayak moved a few inches and then, with a sly slowness, rolled sideways emptying itself of its human burden. Gerry plopped into the water and sent off waves that pushed his empty craft forward, with just enough force to free it from the hold of the beaver dam. While Gerry emptied the water out of his rubber boots, I got out of my own kayak and pushed it over the slippery twigs and branches of the dam. Undaunted, Gerry boarded his craft again and let the sun and vigorous paddling dry his clothes.

A raven was driven by a flock of small birds from a barkless tree that had been gnawed at by many storms. It voiced its displeasure cawing deeply and skillfully circled upward until the attackers fell back. Flocks of birds swooshed low over the reeds, their last meeting place before they started on their journey to the south. The reeds on both sides of the channel stood over four feet high, a perfect cover for a silent approach, as we found out moments later. The channel turned back into the river, the reeds thinned out and mingled with shrubs and trees until the bush stood dense and silent. Gerry was slightly ahead when he suddenly lifted his paddle gingerly out of the water and motioned to me to look ahead. About seventy yards down-river a bull moose stood knee deep in the water with his head submerged up to the ears as if he had lost all interest in the world. But I noticed his ears move about and knew that he was interested to hear

115

what was going on above water. Quietly I lowered my paddle and rested it across my knees, so I could quickly use it again. The kayaks drifted slowly with the lazy current and I wondered whether it was safe to get any closer. The bull's head came up, water sloshed from his mouth and we could hear his molars grind plants that slowly disappeared behind the floppy lips. I remembered an old Indian who insisted that a moose can hear a watch tick over one hundred and twenty yards away. Now the bull stopped moving his jaws and turned his head towards us. He stared at us for a few moments and then resumed his meal, ignoring us except for the occasional glance.

He was a strong, well-antlered animal with a dark almost black coat. On his shoulders was a hump which is a combination of shoulder blades and back bones of the neck. The hind quarters were set lower than the front ones and the short useless looking tail was almost undetectable. The antlers were without any trace of velvet, which is usually shed in the early fall. Like all male members of the deer family, the bull moose loses his antlers after the battle for cows is over. The mating season or rut lasts from September to November and during this time the male moose can be irritated easily. In April the growth of the antlers begins again and lasts all summer.

Leafy vegetation and aquatic plants, such as sedges, horsetails and bur reeds make up his summer food. A healthy bull can easily eat sixty pounds of vegetation per day during the summer. I am sure, while we slowly drifted closer, he downed several pounds within three minutes. Like all cud chewers, the moose usually selects a comfortable place later on where he can regurgitate the food for further chewing.

In the winter the moose feeds on twigs of willow, aspen, maple, birch, and dogwood trees, and sometimes pine trees, spruce and fir trees. When winter is harsh and long and good food is difficult to come by, the moose will also feed on the bark of trees.

The moose is truly built for a life in the wilderness. His long slim legs help him travel through the dense undergrowth of the bush and also in deep snow. His cloven hooves spread widely as he wades the swamps and marshes. He is a powerful swimmer and some moose have been seen miles away from shore during their search for new territory.

His eyesight is poor, but his sense of smell and hearing is extremely honed. His summer coat is short and has no underfur. Often the moose escapes heat and flies by spending most of the day in water or mud. His winter coat is made of long coarse guard hair with

117

a fine woolly underfur.

We had drifted so close to the feeding bull that I could see the sun sparkle in his eyes. After the stems had disappeared into the depth of his stomach he fixed his eyes on us again. His nose extended beyond the lower lip and the nostrils were overhung by skinfolds like the wide sleeves on a Chinese robe. From his

throat dangled a handful of wet hair that looked like a false beard that had slipped off his face.

It was obvious that a decision was forthcoming. We stared at each other for a few moments. Then the mighty bull shook his antlers and snorted. Without losing any of his dignity he straightened his head and slowly waded through the water towards the river's edge. I watched the muscles ripple the shiny hide, then relaxed as he jumped up the embankment and vanished soundlessly into the bush.

We both had heard stories of bulls attacking cars and trains. If a bull's ire is aroused he will move through the bush like a bulldozer gone rampant.

Perhaps this bull had never before encountered

weird creatures like us, looking half humans and half
tree stumps that floated in the water. We were both
grateful to him for keeping his cool and marvelled at
his skill to move through the bush so quietly. Much of
the moose's peaceful behaviour that it displays
through the year, such as caution, the desire to seek
seclusion quickly if an unusual sound is heard, even
timidness, are replaced during the mating season by
restlessness and short-temperedness. This trans-
formation is caused by sexual urges that both sexes
share.

Bulls can often be seen feeding on the same juicy
patches, or travel the same paths within feet of each
other, or even stay together in loosely arranged
groups throughout most of the year. The mating urge
ends such peaceful scenes and during the autumn fights
often erupt and the sounds of scuffles often
reverberate through the bush. Seldom are these fights
fatal to the individuals involved, though many leave
the battle field bleeding or with their antlers
splintered. Young and less mature bulls are quick to
recognize their own inability to win against older
seasoned bulls and seldom get past a mock attempt to
challenge their strength. The fight between mature

bulls is a more serious affair that may last several days before one of them will yield. Occasionally antlers may get locked and if they stay locked will mean the death of both animals.

The females, too, show a more active behavior during the rut. They will announce their presence to suitable males with drawn out wailing calls that are often imitated by hunters to attract a love-sick bull.

After the bull has mated with a cow he will leave her and seek out another one. His affairs eventually end when most cows have conceived.

The cow will carry a calf for eight months and in early June she will give birth to the offspring, that

arrives into the world with its eyes open. The first few days in the life of a calf are spent on wobbly legs; when it moves it does so awkwardly. New-born calves are reddish brown in color, with pale heads and dark-eyed rings. They also lack the long flexible upper lip which develops later on. During the second month of its short life the calf begins to sample greens. Moose are strong swimmers and a cow will teach her offspring this important art of survival very soon.

A healthy cow will stand her ground in defending her calf and there is seldom a bear or a wolf that can force her to abandon her young. Adult moose are preyed upon by few predators and even wolves hunting in groups will seldom attack a moose unless the animal is weakened for some reason.

To some native tribes the moose was most important as it provided them with their livelihood. They fashioned clothes and tents from moose skin, tools from bones and the meat was their main source of food. The moose was thought to have great powers and symbolized a long healthy life to the person in whose dream it appeared.

Though nowadays the moose shares its habitat peacefully with the other creatures of the wild, a legend tells us that this was not always so. There was a time when the moose was good-looking and well-proportioned. It often played pranks on smaller and weaker animals with which it shared the same feeding grounds.

The supernatural Wesukachek often travelled on earth to see if his Indian friends and the creatures he had helped create lived well and enjoyed themselves. He came into a lovely valley where he watched with much displeasure as a moose wilfully trampled on the plants that hare, goat, sheep and deer tried to graze on. Being the strongest animal around, moose ignored the pleas of others. Wesukachek stopped him and to teach him a lesson changed him into an ungainly looking animal. He shortened the creature's neck, so he could no longer reach the plants on the ground in comfort, he put a hump on his back to relieve him of his pride, and to make sure he would never forget the

121

lesson, pulled on his upper lip until it became a long overhanging flap, and attached a bundle of hair to his throat. Ever since the moose lives in peace with other animals and keeps to himself in shame.

Though the moose has never replaced milk cows or horses, attempts have been made to domesticate the animal and one such story is told by a young resident of Southern Ontario whose great-great-grandfather was said to have raised a moose calf that later on, when fully grown, proved to be so tame that he hitched it to a buggy quite regularly and drove to church on Sundays, much to the amazement of his miracle-hungry contemporaries.

The majestic bull moose standing in the river that was lined with trees whose colors sang to the glory of passing time was a sight forever to be enclosed in my heart.

At such a time how can one look at the hills, the sky, the water, the trees, hear the birds in the background and not marvel about the wonders of Nature?

THE TIMBER WOLF

With each trip my skill in handling the kayak grew, the wilderness became more familiar and the desire to spend more time outdoors stronger. Every time I went into the wilderness I ventured into a new reality so different from the one of which human society consisted. Despite sore backs and blisters, despite bruises and the occasional state of exhaustion, despite having to cope with the heat, cold, thirst, hunger, mosquitoes and blackflies, I went into the wilderness eager to make up for the many years that I had spent studying a nature that was caged up behind

123

glass or preserved in formaldehyde. The picture which
emerged was invigorating and deeply rewarding. It
was a picture composed of experiences that changed
my perception of life. Even the rocks presented
themselves as living things. The trees began to look
like old sages who knew a lot, but did not talk.
Through the wilderness, mystery shimmered invitingly like the sun that crept into leaves in
transparent splendor.

As I paddled in the quiet water of a secluded lake
beneath a steep cliff where crooked pine trees
precariously hung on to small ledges, I felt lost in
infinity. Like the perplexed traveller in an ancient

Indian legend, in those moments I found myself
transposed into a world where one is invited to rest for
a while, perhaps even live there for some time,
knowing all along that the wrong word or the wrong
thought will instantly wipe away the magic and the
understanding. In this world all comparisons lost their
human values and even cruelty and kindness became
the throbbing of eternity. The creatures that were
reputed to be vicious, merciless killers and blood-
thirsty beasts, appeared in the light that illuminates
the ever self-sustaining cycle of life.

124

The wolf, for instance, is part of this cycle. For too many centuries we have regarded him outside of it. Misled by inaccurate observations of earlier 'naturalists', we have come to believe their claims that the wolf is 'a vicious, useless beast, a coward that favors human flesh'. No wonder that men have put a bounty on the wolf, starting on this continent in 1630 in the State of Massachusetts, 'to get rid of a wasteful and useless predator'. Over three hundred years later, upon taking wildlife inventory, man must confess to a skill that invites the same attributes he once attached to the beastly wolf. Confronted with a lethal array of traps, guns, snares, poisoned baits and chased to death

by snowmobiles and helicopters, the wolf had to retreat to remote pockets of wilderness areas of the northern hemisphere and even there his peace is threatened.

One of the last sanctuaries where the grey or timber wolf still roams in his ageless fashion is Ontario's Algonquin Provincial Park. In the north-western corner is the little village of Kiosk whose houses gather loosely on the shore of Lake Kioshkokw like a disturbed herd of deer that came there to drink. Neighboring as close as two wingbeats of a heron are three more lakes making this area an ideal set-up for a boat trip. The country is as rugged and as wild as it must have been thousands of years ago when it only knew the silent footsteps of Indians. The bush is so dense that it quickly drowns out any sound from the nearby village and even the hum of motorboats is swallowed up before it can travel too far.

One summer, I paddled quietly through the afternoon light close to shore when I glimpsed what I thought was a German Shepherd dog. The animal stood propped up with its front legs on a log and peered at me curiously. I called the dog and asked it to remain there to share my lunch. When I drew the kayak nearer to see if it was wearing a collar, it disappeared. I paddled on hoping to find a spot that was good for camping and swimming and moments later had the weird feeling of being watched.

Through the shore foliage loomed the dark figure of the same animal that stood as quietly as before—watching me. This time I did not approach it directly, but paddled about a hundred yards past it. I planned to tie the kayak to a tree and walk back along the shore in the hope that I would surprise the animal. When I turned shoreward well past the spot where I had seen the dark figure for the second time, a head suddenly popped up from behind a rock. In the bright light of the afternoon sun I saw that it was the head of a wolf. Its ears stood straight up, intently listening to every sound I made, and its eyes were transfixed on me. With slow short strokes I manoeuvred the boat closer until the wolf became uneasy. It ducked down,

127

but uncertain of an early retreat stuck its head up again, keeping itself in readiness for a quick getaway.

Its cheeks were covered with short pale yellow hair that lightened to a white patch on the chin and throat. Greyish brown hair darkened the upper half of the head and outlined the ears. The coat was well groomed and gave it the appearance of a strong and healthy animal. The eyebrows were slightly drawn forward to shield the amber eyes from the sun. Its steady look reflected concentration, confidence and curiosity.

Another stroke shortened the distance between us beyond the limits of tolerance and in a flash the rock was barren and the wolf had vanished into the shadows of the bush. For two miles the wolf played the game of 'Come closer if I let you'. The last time I saw the wolf, it stood in full view in a small clearing that had a backdrop of blueberry bushes and sumac shrubs that thickened into a dense curtain of undergrowth. Like the hull of an overturned boat, a granite formation extended from the grass into the water in a gradual decline. The wolf seemed to say: "This is the place you have been looking for." Slowly and carefully I drew closer until I was about thirty feet (10m) away where I stopped, then sat motionless so we could look at each other. For a good while, neither of us made a move. Like an expensive woollen blanket the dark gray coat of fur was draped over its head, neck, shoulders and back. From there it extended to its tail. On the throat the gray lightened and ran down to his chest where it turned into white that covered the underside and legs. On the legs was a hue of rusty brown. The coat was sprinkled with patches of white and rust that blended into the dark gray like patches of sunlight diffused by foliage. The wolf stood like a well sculptured monument, with its front legs together and one of the rear legs a half step ahead of the other. The tail was pointed low, the neck was stretched forward and the head was turned sideways to face me.

I could not help but deplore the fact that this beautiful creature was regarded as an enemy of man,

128

while the dog which descended from the Asiatic wolf
has become man's best friend. Yet the wolf is nothing
but a wild dog, that has all the qualities we admire so
much in a dog, except that it likes to remain
independent from man.

Wolves are sociable animals that live in packs.
These are made up of the parents, offspring and
relatives. Hunting is a collective effort. Usually when
a large animal such as a deer, is selected as prey, the
hunt is done in relays. One member of the pack starts
the chase and runs until it is tired. Another member
takes over until the deer will either slow down or
another change is necessary. This goes on until the
victim is exhausted and can be pulled down. A single
hunt often takes a pack sixty miles or more.

Studies have shown that wolves prey on sick or
weakened animals and thus are important to the
balance of nature. Wolves are quick to determine a
weakness in their prey, but how they do this is still a
mystery, as yet unsolved by today's biologists. Wolves
also have an ideal family life.

When a male is three years old he will look for a
mate. Once paired, the couple remains together for
life. They select a den or may dig one together. In
April or May the den is filled with the squeals of blind
pups that average six in number. According to a game

guide, wolves have their own birth control system. This man believes that parents will allow their pack to grow in size only up to a certain number. If the number of pups born surpass the allowable number of family members, the excess pups are killed off immediately after birth. If enough deaths among the older members have occurred before birth, then all newborn pups are allowed to live.

The fuzzy little creatures are unable to see their immediate world for ten or twelve days after birth. Initially the mother provides the pups with all their nutritional requirements by suckling them. Later when solid food is added to the diet, the father gets involved. He is a model provider and though he snoozes mostly during the day on a lookout near the den, he stays tuned in to the world with his ears. He hunts mostly during the night and returns to the den every morning with a supply of food. Eventually the pups are allowed to romp in front of the den and at the age of twelve weeks are moved to a new home. The new hunting grounds serve as a classroom where the youngsters are taught the secrets of survival in the wilderness.

Wolves, like their relatives, the coyotes and foxes, are meat eaters and have strong jaws and well developed teeth that are necessary to tear meat and crush bones. Mice, squirrels, groundhogs, and rabbits are pr :ferred to larger animals, but deer, elk, caribou, and moose are preyed upon as well—especially in winter.

Much of the wolf's reputation is a product of human fear based on myth instead of fact. I am sure that many campers and hikers would never venture into certain wilderness areas if they knew how closely they would be watched by wolves. I became aware of their silent presence many times during my wilderness trips. Once my companion and I had spent the night sleeping out on a gravel bank in the middle of a river and as we crossed over to the sandy shore we saw wolf tracks that indicated the animal had satisfied its curiosity about us and wandered off without so much as scratching us.

130

At another time we camped near a spot where a group of hunters had gutted six deer whose piled-up entrails were still warm as we pitched our tent. It began to snow lightly before darkness set in and the next morning the guts were gone. Tracks of wolves were all around our tent and criss-crossed the woods around. We followed a heavily travelled trail for three miles and never saw a wolf, yet the next night we could hear them howl nearby.

It is this howl that has always been described as 'blood-thirsty', yet it has nothing to do with the animals' feeding habits.

Like the screech of the owl, the cry of the raven, or the honking of geese, the howl of wolves is as much a part of the true wilderness as the roar of a rapid or the sky-jabbing growth of a cedar tree. There are many who would like to see the wolf exterminated, but if we question its right to live in the wilderness, how long will it be before we stop the bluejay from migrating south, or the otter from eating fish?

To many Indian tribes the wolf played a prominent part in ceremonies and was often represented on crests, totem poles, rattles, masks, and other ceremonial objects. Great magical powers were

attributed to the wolf and many legends tell of his ability to change form, especially into the form of Killer Whale. As such, the wolf could make both the land and the sea one vast hunting ground. In some legends the wolf exercised gratitude for kind acts by people.

Once, a large wolf appeared outside a village and though people were afraid of him they invited him into their communal building where their Chief offered him some food. The wolf never touched it and the people were certain that something was wrong with the animal, so they examined him. A large bone was found stuck in the throat and removed. After, the wolf was invited to remain in the village and treated to an elaborate meal but the animal vanished overnight. Several years later, during winter, the tribe experienced difficulties that brought the people close to starvation. Just then a mysterious stranger appeared who beckoned some of the stronger people to follow him into the bush. He led them to a food cache that was filled with the meat of various game animals. Before they could thank the stranger for their salvation he had changed into a huge wolf and trotted off into the wilderness.

THE STRIPED SKUNK

Herb was a jovial fellow whose fondness for beer was visible in the form of a pot belly. He was only twenty years old when I met him during the first summer of my new life in Canada.

Of medium height, Herb possessed a voice that commanded the attention of people who would normally have been out of earshot. He had a sense of humor that centered around the basic things of life,

He was good-natured and always ready to do things in a manner that lacked responsibility. Shortly before I met him he had grown a beard in the hope that it would make him look older. But the following events made it clear that maturity is a by-product of experience well taken, and not a matter of looks.

Herb's string of unlucky events began during a fishing trip he made with an Indian friend. After several attempts to hook a fish he suddenly lost his temper and flung the poor frog against a rock.

"You should not have done that," the Indian scolded him. "Don't you know if you mistreat a frog you will have bad luck?"

Herb only replied with a series of grunts, of which the most civilized one sounded like 'shut up'.

The little frog lying there with its belly split open did not seem capable of inflicting bad luck on anyone and to prove the point Herb went downtown and bought a fibreglass boat with a big engine for twenty-five hundred borrowed dollars. Four days later the engine gave up its life, just as Herb was skimming the surface of Trout Lake. It had completely burnt out and left him helplessly drifting, too far from shore to swim for help. Hours later he was taken in tow by another boater who realized something was wrong when he saw Herb waving frantically. Back on shore Herb hitch-hiked home, borrowed his father's pick-up truck and promptly backed it into the lake while trying to load the boat. The costly bill to recover the truck and repair the engine should have told him that perhaps the power of the little frog was far greater than he cared to admit. But he saw no reason to repent.

A few days later Herb decided to paint his father's summer cabin. The cabin should have had a new coat of paint a long time ago and armed with paint, brushes, rollers, ladders, and plenty of good will, Herb set to work, to beautify the place. A few hours later he turned up at the neighbour's cabin panting heavily, eyes crazed with panic. All they could get out of him was the word "fire". By the time the fire fighters arrived there was not much else for them to do but to make sure the fire would not spread to the surrounding bush. Herb, with tears in his eyes, insisted that he did not know how the fire got started.

In the back of his father's lakeside property stood an old tool shack that no one had used for a long time. A few days after the fire Herb spent a weekend with the help of some friends clearing the site for a new building. It was decided the shack should be ripped down too. Herb armed himself with an axe, I grabbed a crowbar and together we began to demolish the shack. Unknown to us, a family of skunks had claimed the space under the floor as their home. Angered by the noise and the prospect of loosing the roof over

their heads, one of the parents darted out from under to investigate the situation, or perhaps only to lead his mate and offspring to quieter premises. Flitting back and forth with its tail raised and stomping the ground like a mixed-up dancer it made its irritation known.

I had never seen a skunk before and asked Herb what the name of the animal was that had a long bushy tail, a white stripe on each side of its back and

136

behaved a bit nervously, but he was too busy smashing boards. He was so totally given to the chance to work off his frustrations that he neither heard me, nor saw the skunk, as he stepped off the rubble between the excited animal and its family. That move quickly brought the worst out in the skunk. With amazing speed he doubled up sideways, so that his front and hind legs were lined up in a row, like two different animals standing side by side, front and rear together. From the glands near the anus, that looked like the evil eye of a monster, it squirted a rancid liquid at Herb's legs.

For some time after Herb avoided his friends completely, or rather they avoided him, but eventually he went to Toronto to start a new life. His father told me later that Herb was very happy there.

The foul odor of the striped skunk once smelled is an experience that no nose will ever forget, nor is it easy to make the offending smell go away. The best remedy still seems to be time. Unfortunately, it takes weeks, even months for the odor to vanish. I first noticed this when I drove a borrowed van along the highway to Martin River. While following another car I failed to avoid the carcass of a skunk and drove over the dead animal with one wheel. For weeks after, the smell was so strong that the owner could not park the van in front of his house. When it finally seemed to have gone it returned with every rain shower for over a year.

The scent of the skunk is part of an oil substance secreted by two glands which are located on either side of the anus. One tablespoon of the musk will last for four or five discharges which the striped skunk resorts to with great reluctance. The skunk gives a number of warning signals, one of which is a handstand. Though it walks back and forth on its front legs with its tail raised it does not spray from this position, but has to lower its body and shape it into a U-form, so that its head and rear are side by side. The spray is accurately aimed and will reach an object as far as twenty feet (7m) away. The skunk's real danger to other animals and to man lies in the disease of rabies of which it is a major carrier.

Belonging to the weasel family, the skunk is a most useful animal that inhabits the mixed forests and grasslands of North America. It roams in the open areas and searches for insects, mice, squirrels and other small mammals, as well as birds, and even plants. Sometimes a skunk is the source of annoyance to a

farmer when it raids a hen house or a bee hive, but in
general the animal does its best to keep the insect
population from taking over the world.

Skunks spend the winter in communal dormi-
tories underground and by February or March emerge
from the dens to breed. A litter of six naked and blind
young ones usually arrives at the beginning of May.
Two to three weeks later they are able to look at the
world with their own eyes and are dressed in their
characteristic black and white silken fur coats.

Hawks, owls, and lynx are some of the few
predators that will risk getting a new body odor by
preying on skunks. A much greater hazard to skunk is
the motorist, because the animal will not forsake its
confidence in its offensive musk when crossing a
highway.

In Indian folklore the skunk appears as a simple
fellow who has great powers but does not know how to
use them to his advantage and thus become the source
of good-natured humor.

At one time a pretty maiden was lost in the forest
and found by the skunk's old mother who offered her
shelter and food. When Skunk came home and saw the
girl he fell in love with her, but the girl remained cool
because of his powerful scent. During the night the
girl sneaked away and ran through the forest until she
came to a wide river which she was unable to cross.
She climbed the tallest tree to rest in safety. Soon
Skunk appeared following her tracks. Thinking that
she had swam across the river he was about to plunge
himself into the water when he noticed the girl's face
deep in the water which he did not recognize as a
reflection. Immediately he took a deep breath and
dove into the water to bring up the girl. His
underwater search revealed nothing and he came up
gasping for air. Looking into the water he saw the
smiling girl beckoning him to come for her. Again and
again he dove to search the bottom of the river and
always came up empty handed. Yet each time when
the water had calmed he saw the girl deep down,
smiling. Dripping wet and shivering with cold he built
a fire to warm himself after each dive. Sparing no
effort he dove into the river until he was barely able to
come back up for air. He pulled himself up onto the
land and collapsed into the flames of his fire. His
dripping fur extinguished the flames and no harm
came to him but the ashes blackened his coat except
where he came to lie on unburnt logs. This is why, to
this day, the skunk has a black coat with white stripes.

140

THE BLACK BEAR

Near the village of Spanish, Highway No. 17 slices through an Indian reserve where a stocky, five-foot-tall Indian named Joe lived. He was the prospective father-in-law of my friend's brother. One day my friend and I drove there to meet the family. Joe supported his wife and three children by trapping, fishing, and hunting. The proof of his skill was everywhere. The freezer was stuffed with cuts of moose meat, rabbits, an assortment of fish, and pieces of bear meat. In a big pot on the stove simmered beaver tail soup and outside, the enormous hide of a black bear, complete with skull and claw-bearing paws, hung stretched between two birch trees.

Joe wore an apologetic Santa Claus smile as he showed us his single shot .22 calibre rifle which he used for hunting.

There was hardly any varnish left on the stock of the gun, the barrel was tied down with parcel twine and the whole thing looked just about right to be used as a walking stick, rather than a rifle for killing wild

animals. "Very old gun," Joe mumbled, "very, very old, but good!"

Maybe so, but I would rather see it in a museum than use it for shooting a black bear with one bullet, which is exactly what Joe had done.

"Him know everything," Joe went on pointing at the bear's head that swayed slowly in the wind. "Him smartest animal in bush." He nodded his head to emphasize his conviction and sat down on a pile of firewood to tell us his thoughts on the black bear.

It all started at the time when plants, trees, rocks, rivers, lakes, and creatures spoke the same language, and were capable of changing appearances. During this time of universal understanding a state of collective beauty existed and the beings of the earth

142

were kind and showered one another with compassion. The happy carefree time eventually came to an end when bad spirits invaded the earth and spread their evil thoughts among the earth's creatures, some of which became as wicked as the spirits themselves. During the resulting confusion, the bear remained true to himself and steadfast in his love for his fellow beings and thus earned himself a special place amongst all creatures for all times. To this day he is gifted with a wisdom which surpasses that of all other animals and most humans.

Most people mistake his clumsy, comical behaviour for lack of intelligence, but some of the Indian people know better.

"We know him talk to spirits a lot," Joe confided.

143

He added that one can often see a bear stand upright among the trees, seemingly engrossed in a conversation with invisible beings. His head will move from side to side, his lips will utter soundless words and he will trust his ears at the same time as if trying to listen in on a conversation between several people.

The special relationship which the bear retains with the spirits that control the earth and the universe entitles him to much reverence and respect. Before one attempts to kill a bear one has to empty the mind of bad thoughts. When meeting the bear he has to be addressed with the respect befitting a great ancestor and offered an apology for the act of killing. He has to

144

be assured that all his bones will be returned to the ground unbroken and that not the smallest morsel will be wasted. He has to be asked that his strength and wisdom may be imparted to the hunter and thanked for allowing himself to become the food of the hunter and his family. Then the bear will know that he is to perform a good deed again and will keep still to be shot. "Any gun that shoot straight will do, and one bullet is plenty to kill the bear". Such was Joe's conclusion which sounded quite credible.

One would expect that the ideal place to see black bears is the northern wilderness as they are widely distributed throughout the forested regions of Canada. Yet they are difficult to observe because their acute sense of smell tells them from great distances when a man approaches. On the other hand it is often easy to see them on the fringes of human habitation where they are attracted by the odor of food or food wastes. During the years I lived in North Bay, a number of black bears were shot by people whose garbage cans they raided. The bears are not only lured to the back porches of houses by food smells, but also to campsites or garbage dumps where they can often be seen in groups.

Bears have a wide range of wilderness food to choose from, including nuts, roots, grasses, berries, wild honey, insects, birds, fish, small mammals and occasionally the carcass of a larger animal that either died of disease, injuries, or is the leftover of a kill by predators.

Throughout the summer the bear is engaged in a long feast and by fall its fat body is dressed in a shiny coat of fur that is the sign of excellent health. When winter is near the bear can predict the first severe snowfall and will retire to a cave or a huge uprooted tree that gives it enough shelter to curl up for the sleep that lasts all winter. This winter sleep is often mistakingly referred to as hibernation. In true hibernation the body temperature, heartbeat and respiration are almost undetectable. When an animal such as the bear, skunk or the racoon spends the winter in a deep sleep its body temperature and

145

heartbeat remain normal, and breathing is only slightly slower. A hibernating animal cannot be aroused when disturbed, but a bear found during its winter sleep may be awakened easily. A bear may wake up occasionally and stumble around and though it may look sleepy, it is better to stay out of its way.

During the winter sleep the female gives birth to two or three cubs that arrive in January or February. They are extremely small, about the size of squirrels, blind and toothless. The mother will nurse them and keep them warm throughout the winter sleep. Sometime during March the family may make short visits to the outside world until they have all shed their grogginess. The cubs are nursed well into the summer and gradually weaned. They are brought up by the female alone and aside from the mother and cub relationship and the short courtship between the sexually mature, bears prefer to live alone.

The cubs undergo extensive training in wilderness survival until their mother chases them away in favor of a new generation. Since bears breed every two summers the cubs' age when they have to start a life on their own is somewhere between eighteen and twenty-four months. Though the female is a strict disciplinarian, she is also a very tolerant and concerned mother.

Just like children, bear cubs are always full of high-spirited fun, which will sometimes get them into trouble. With whining, babylike cries they call mother to the rescue. Security, love, joy, and happiness all find their expressions on a cub's face as it snuggles up to its mother at nap time or as it comes for a quick cuddle before it romps off, driven by the same curiosity that make children explore every nook and cranny.

During various boat trips I came across signs that hinted at the presence of bears. However the first opportunity to observe a bear came not while I was out in the wilds, but while I was comfortably seated in a deep armchair. I was visiting a farmer near Powassan, Ontario, and we were having coffee and buns. The conversation was filled with personal tidbits when the door burst open and the farmer's eight year old

146

daughter forced her younger sister into the room. The five-year-old girl was crying because she wanted to play with the cute teddy bear that sat under the apple tree. The furry playmate turned out to be a black bear cub carefully watched by its mother from the edge of the garden. She was standing upright and leaning with her elbows on the top pole of a wooden fence with an expression similar to a station master watching a train leave.

The cub sat under the tree unsure of what to do, looking at mother for a hint. But mother was not yet willing to give any instructions. When she spotted us she whined to summon her other offspring. For a while she stood there sniffing the air and listened behind her where we assumed the other cub was. Indeed, shortly after, the second cub came galloping into view across the field, slowed down next to mother, crawled through the fence and pounced on its playmate. They wrestled for a few moments then slapped at each other and in an abrupt fit of passion embraced each other. They sank to the ground holding each other, then rolled apart. With unexpected quickness the mother hurdled over the fence and with the typical shuffling gait walked over to the tree. The cubs sat up and

147

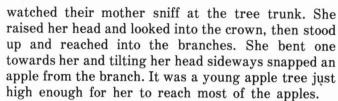

watched their mother sniff at the tree trunk. She raised her head and looked into the crown, then stood up and reached into the branches. She bent one towards her and tilting her head sideways snapped an apple from the branch. It was a young apple tree just high enough for her to reach most of the apples.

Too busy with their farm work, the people had not had time to harvest the orchard and some of the over-ripe apples dropped to the ground almost into the laps of the waiting cubs. The cubs lay down like dogs when they chew on bones and sank their teeth into the juicy fruit. The mother looked at the cubs and decided that she, too, was more comfortable in a sitting position. Slowly she sank down on her behind while she kept her back straight like a typist and munched on an apple.

For over one hour we watched the bears satisfy their fondness for apples and play. All too often we thought we saw in some of their movements the movements of people, and I kept hearing Joe as he said, "A bear is one of us."

The family of black bears departed so abruptly that we thought perhaps they could no longer stand to see us watch them. Suddenly the mother stood up, cocked her ears forward and listened. She sniffed the air, then dropped back on all four feet, emitted a drawn out "woof" that jerked the cubs out of their playful mood and sent them running past the house towards the nearby forest. She followed them without hurry. A few minutes later a pick-up truck came huffing and puffing down the road trailing a cloud of dust. It turned into the lane that ran up to the house.

Bears are not only good runners, but also excellent climbers and will climb any tree that is strong enough to support their four or five hundred pounds. They are also good swimmers and once I watched a black bear cross an inlet on the west coast of Vancouver Island.

Black bears are not always black, often they are shades of brown, cinnamon, or even a bleached blond. Some of them have white chest patches.

The bear has only one major enemy: man.

148

THE RIVER OTTER

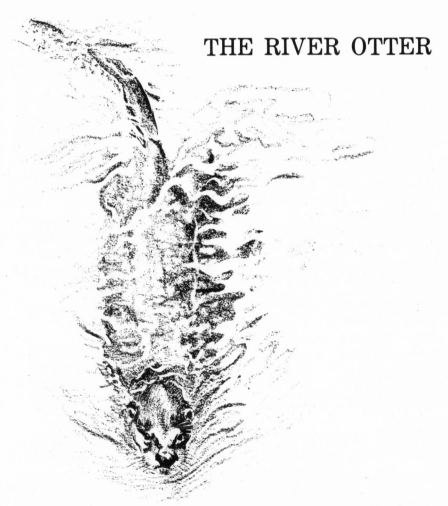

Since man's appearance on earth, rivers have invited him to share the land they keep fertile. We owe much to the rivers that have allowed many a race of people to grow from infancy into a highly developed nation, whose huge stone cities now drown out the chatter of small waves and the song of swirls and eddies. Rivers have made nation-wide travel possible

149

and, blocked by dams, have given modern man warmth and light. Sadly, too many of the rivers are sick and it seems the sicker the river, the richer the nation that holds the water rights to it. Some rivers have already died and become a flowing mass of sewage, grease, and chemicals. One such river near Cleveland one day caught fire, which could not be dowsed with its own 'water'.

The fate of too many rivers is mapped out by politicians and industrial engineers and so, more and more they lose the freedom to flow on their own accord.

In the great northern wilderness one can still find rivers that freely bestow on man the greatest gift they

have to offer: solitude.

The wisdom of a wild river has often been contemplated by thoughtful men who admired its steadfast flow that brings it to the ocean, its patience as it winds slowly and with great perseverance through the land seeking the least resistance, its skill in evading obstacles it can neither move nor carve away, and its serenity in seeking the lowest level for its travel.

To the Indians a river was very important. They thought it was alive with a female divinity—The Creek Woman—and in each brook or creek this gentle spirit gave life or death to the beings that lived in and nearby the clear streams. She ruled over the fish, the

150

frogs, the birds, the trees, and all the other creatures that sometimes came to drink. The abode of each Creek Woman was at the source of the river and though she had great powers she seldom used them to the detriment of the river people. Mostly she saw to it that the river flowed in peace.

Once I paddled up a narrow wilderness river until I reached its source: a little lake called Clear Lake that was no bigger than a man-made fish pond. In a silent, serene mood the trees surrounded it and cast their dark shadows onto the water for most the of the day. But at high noon the surface came alive in the glitter of the sun and the bottom of the lake shone through the emerald color of the water. Below, numerous springs bubbled through the sand where rocks and trees lay as ghost-like shifting shadows. Sometimes the shadows moved in human forms and I was sure I had discovered the abode of the Creek Woman. I stayed by the lake for four days and listened to its moods that made me feel as if the universe stretched right through me. The Creek Woman showed her kindness and commanded some of her creatures to visit me.

First, I heard the squeaks of an otter as I sat inside my translucent tent taking notes. When I looked up I saw the face of an otter outside the plastic with an expression of "who are you?" It stared at me for some time and since I did not move, it came closer. It braced itself against the plastic and peered inside like a child that looks through a shop window. Long whiskers sprouted from the upper lip like an unkempt moustache and a few drops of water hung in them like dew drops on grass. Its entire fur was sleek and wet and showed the contours of a well-fed animal. The small eyes squinted a couple of times to make sure that what it saw was still there and not some kind of mirage. Then it moved its head sideways and looked at me from its left eye. The eye was nearer the nose than the small round ear, and the color of its cheek was pale grey. Down the chest a golden brown was strikingly set off against the dark brown fur on the flanks and back, while a few long hairs extended from the eyebrow. The otter faced me again and braced

151

itself with one paw against one of the tent poles and
with the other one clutched its chest. The nostrils
alternately widened and narrowed. Again the animal
leaned forward and pressed its mouth against the
plastic and tried to chew it. Then it pushed its head
away from the tent and listened to what was going on
behind it. Both front paws were bulging the plastic
and showed the strong, webbed toes with their sharp
claws. Then something else seemed more interesting
and the serious examination was over. The otter
turned with a dachshund-like movement, swooshed its
tail against the plastic and hobbled to the water.

Otters are actually large aquatic weasels that fish, travel and play in the water, but reside on land. Their well defined trails run alongside rivers and lakes and evidence of their presence can be found in the form of droppings that will usually have fish or crab remains in them.

Otters feed mainly on fish, which they catch in swift pursuit. Foraging along the shores, they will also feed on birds and small mammals. In a den under rocks or roots, or in a vacant beaver lodge, up to half a dozen young ones are born during March or April. When the young are about three months old, the whole family ventures out to fish and since they are very fun loving creatures, they constantly mix pleasure with serious food gathering activities. There is one pleasure otters love above all others: sliding. Wherever they are, they always look out for a mud puddle or a steep clay bank to slide down, often spending hours in plunging themselves into the water from the slide. Even during the winter, otters will not forsake this addiction. They can often be seen pushing themselves with their hind legs, while their front legs are folded over their chest or held sideways so that they slide on their chests and bellies through the snow.

Otters not only dive effortlessly, but can swim short or long distances with great speed and may stay below the surface for minutes at a time. An otter's webbed feet can quickly propel it through the water while its body and tail assist it with a unique sequence of flexes. Sometimes these undulations are employed without the use of the feet, especially during playtime.

Less than a quarter of the otter's four-foot long (130cm) body is made up of the tapered tail, which is a strong support for its many activities. While swimming, it serves as a rudder, on land it is used as a

153

prop, sometimes as a weapon, or a tool.

Behind the flat forehead lies the centre of all the quick reflexes the otter has: its drive for motion and grace, its endurance in travelling and its courage that will make it hold its ground against coyotes, wolves, wildcats, and foxes.

For a family of otters life is a series of tags, pushes, rolls, flips, jumps—all done in good fun. A kayak that is half in water and half on land, has a hollow inside and a slippery top offers fun no otter family can ignore.

Heads kept popping through the water around my kayak so often that I was unable to tell how many there were. One of them climbed out of the water and sniffed at the kayak. It climbed on top of it, slid off and

stumbled down like a clumsy tightrope walker. It came to lie on the grass on its back acquiring a new view that it greeted with a series of chirps. Another one poked its head out of the water, looked at another otter that was lost in contemplating the sky and decided to return the dreamer to reality by trampling over it. The offender was quickly chased into the kayak where both squealed and thumped about. Three more otters left the water to see what was going on inside the kayak, but now they were crowding each other and, recognizing a slippery surface when they saw one, the scrambling-about game turned into a 'push, slide and fall into the water' event.

The pastime lasted for several hours judging by the sun's journey. The delightful time went by all too

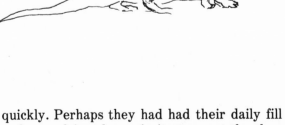

quickly. Perhaps they had had their daily fill of play and were driven by curiosity to scout the shores down river, or perhaps the simple urge to eat sent them in search of food. By midday they felt compelled to leave. While they scrambled over rocks and fallen trees they gradually disappeared. I sat for some time by the water where they had played earlier, hoping they would return to again display their grace and ease of movement, their effortless control over their bodies, their delight in play and games. I wondered if I would ever again have the chance to hear them squeal, chirp,

grunt, and chuckle as they wrestled and raced each other in the water or on land.

Like most animals of the wilderness the otter, too, is being decimated and man has fewer and fewer chances to see them wild and free. Only a fraction of the once hundreds of thousands still roam the wilds of the North American continent.

To make a fur coat, six otters have to give their lives and their pelts. Uncounted baby otters are thus forever deprived of their chance to live. More power dams, more highways, more polluted rivers, more insecticides see to it that fewer and fewer otters survive. For every light bulb we turn on, the sunlight that shines for the otters grows just a little dimmer.

156

The Indians regarded the land otter as a sacred animal and never ate its meat. The spirit of the land otter was most benevolent to Indian women, and most of them sought to acquire it. Any woman who possessed more than one otter spirit also possessed great respect within her community. The otter spirit was harmful at times and especially men avoided it. If a man drowned, his body was caught by the spirits of land otters that lived in caves, sometimes under water—transforming him into an otter. This changed man would in turn approach a single relative who was out hunting or fishing and try to lure him into the den of the otter spirits, which would mean certain death by drowning for the lone hunter.

Once, four brothers went out in a canoe to fish, but were surprised by a storm that overturned their craft. Another canoe appeared and the brothers were rescued by its occupants which they believed to be their relatives. They were taken to shore where they realized that the rescuers were land otters disguised as relatives. Then they knew they had actually drowned.

The empty canoe was found by their real relatives who tried to free the brothers from the land of the otters. They killed many otters, but the people suffered many deaths and eventually tired of the long war. Then two white otters appeared in the village of the Indians who recognized them as messengers from the otter chief. These spirits were treated like royal guests and they reported back to the chief that the Indians no longer wanted to make war on them. Ever since, land otters have refrained from luring people into their dens, changing only those that drown by accident.

157

THE LOON

Somehow the otters seemed to have taken the friendly and joyful mood with them when they departed from Clear Lake. I heard their last chuckle in my ears for sometime, and when it too subsided, a sadness fell over me like a fine meshed net that is almost invisible, but hinders every movement. The quietness of the wilderness became oppressive and the vastness of the land make me aware of my own smallness. Once I heard some birds twitter through the trees, otherwise everything was as silent as a grave. Even the wind seemed to have left for distant lands. Only the sun showed some life. It burnished the ground with rays that fell in handfuls on the forest ground where they faintly trembled. Around me the trees stood solemnly like a huge crowd waiting to be addressed during a funeral, but the only voice that spoke was in my mind. It told me of the loneliness that befalls those who are unable to share beautiful times with loved ones. It deplored the fact that one can learn such a lot in a lifetime, but to enjoy life one needs to move in the spirit of love. Then our follies are forgiveable, our joys limitless and our wisdom of true value to others.

Slowly the shadows shifted and told me of the passing of time. Evening came. The wind returned, gently caressing the land. The trees whispered, birds sang and a chipmunk scurried through the grass like the spirit of sunset. I heard the wings of a large bird beat the air as it flew into the darkness at the opposite side of the pond, where the night had already settled in. There was the splash of a body landing on the water that foamed where the bird had touched down and a few moments later, small waves sloshed against rocks and tree trunks that lay scattered on the shore. Enchanted by the sunset I felt less lonely and as I kindled the evening fire, I thought of the millions of people who never know such tranquillity.

From the darkness, where the bird had landed, a whimpering arose that sounded like the soft cries of a small child that had been cast out by loveless parents. The wailing climbed higher until it hovered over the tree tops, demanding the attention of forest dwellers from near and far. The high pitched scream changed into senseless laughter that is so often indicative of a mind lost in deep grief. Indeed, as if the bird had to bemoan some intolerable injustice it started anew, haunting the evening with a mood of desperation. I knew then that I was listening to a loon, though I had never seen one before nor heard its moanful call. My first Indian friend, Gerry Dokis, had told me enough about the loon so I was able to identify it by its call. He also told me the legend that explains why the loon has such a wailing cry.

A long time ago, there lived a brave young hunter

who was much liked by his people. He fell in love with a beautiful princess and everybody wished them well, except for a powerful and evil shaman who also wanted her for his wife. Unable to watch the joy of the lovers any longer, he once followed the princess to a river where she was to take a bath. Overwhelmed by her splendid beauty he pleaded with her to become his wife. The princess only rebuffed him with disdain which angered him so much that he cursed her with a spell that changed her into a loon. Still not satisfied, he moved her legs back. Then he told her that she was to find her food in the depths of lakes, to make sure she would stay far away from her lover. This is why the loon walks only with great difficulty and spends most of its time diving. During the quiet evening hours when lovers usually meet and loneliness, despair and sadness are hardest to cope with, the princess, in her disguise is no longer able to control the longing for her lover. With heartbreaking cries she calls for him, because only he can free her from the curse.

The loon wailed and cried all evening. In between calls it listened momentarily to the quietness as if awaiting a response. When the last glimmer of daylight faded, the loon ended its call with low giggles that saddened me again for they seemed to convey despair.

Down by the river a frog grunted until others responded. The shore came alive with the sounds of countless cello and bass players practising on untuned instruments. The noise increased as the darkness grew deeper. Stars etched themselves into the blackness and the Milky Way blinked its message onto the surface of Clear Lake, where the loon drifted while dreaming of happier days.

I retired into my sleeping bag and by the dim glow of a flashlight consulted a guidebook on birds to read up on the common loon, which like so many animals in our vanishing wilderness is harder and harder to find.

During the winter months the loon's plumage is a drab gray that seems to blend into the cold unfriendly saltwater where it fishes. Come spring the birds migrate inland to find fresh water lakes where they build their nests, just inches above the waterline. Nest building is done purely out of necessity and lacks design. It may be built on a rock, a few feet from the shore, or on a pile of floating debris that became anchored in some reeds, or fallen trees. It is usually a maze of sticks and some mud, the top is flat and has a very shallow dent, preventing two gray eggs from rolling off. While one loon sits on them, always ready to glide into the water at the slightest disturbance, its mate keeps guard, ready to lure a predator away.

Sometime in June two young loons are born and a summer of teaching follows. Though the young swim

without lessons, they cannot dive without receiving proper instructions and lots of encouragement from their parents. Success comes slowly and first the little feather balls find it hard to overcome their buoyancy, but eventually their efforts are rewarded and they manage to stay under water for a few moments. Within two months they are as skillful in diving as their parents. A seldom, but always heart-warming sight, is a loon mother drawing a silent wake through the still surface of a lonesome lake, while her two young ones hitch a cozy ride on her back.

A loon is a superb diver and if threatened will swiftly vanish below the surface leaving but a momentary ring to tell of its disappearance. Under water the bird swims with great skill, propelling its streamlined body with its webbed feet, sometimes aiding them with its wings, and is able to stay submerged up to three minutes. When it reappears it will often stand up in the water and shake the wetness out of its wings before vanishing again.

On land the loon's skillfulness is limited to froglike hops and an awkward shuffle, because its legs are not easily able to balance the body's weight.

During the spring the loon's color changes to a deep black that seems to be coated with a green varnish. On its back is a "checker board" and, draped around its neck and throat, a collar that looks like a necklace of white pearls. According to an Indian legend this necklace was given to the loon by an old blind Indian in gratitude for restoring the man's sight.

When afloat the loon sits low in the water like an overloaded boat, and displays alertness as it relentlessly pursues fish, which it catches with its strong beak during speedy underwater chases. It takes to the air only after running on the surface of the water a great stretch while wildly flapping its wings.

Loons are devoted partners and an old wilderness loving lady who witnessed the return of a loon pair every spring for four decades insisted that they remain true to each other even after one of the partners dies.

THE LYNX

Next morning, while the fingers of the first sunlight reached into the cloudy sky, I dressed and went quietly to the water where I made myself comfortable in an old tree stump that looked like a primitive armchair. With my binoculars I searched for the loon that seemed to have vanished.

On the far side the shore was faintly visible through a pale bluish-green mist that, ruffled by warmer air, drifted upward in small curls. A handful of sunshine dropped through a hole in the clouds and played on the aged bark of a weathered pine, while a lazy wind stirred in the tree tops.

Guiding the binoculars along the shoreline I saw nothing but rocks, bushes and a maze of branches;

some withered, some full of life, but all grappling for air and light. The contorted trunks of tree skeletons could be seen above the water and the remains of a beaver dam lay heaped up near the spot where the water of the lake spilled into the river bed. Fifty yards further down, the river turned sharply and disappeared behind a wall of trees.

I lowered my binoculars and leaned back to savor another breath of the morning air that was so deliciously scented with freshness. Then I saw a movement near the beaver dam. Slowly I raised my binoculars again and after focusing spotted the head of a lynx. With stiff ears the cat listened, keeping its neck stretched and its head as motionless as a sculptured bust. A tuft of black hair sprouted from each ear tip and the cat's gray-brown face was framed by white tipped mutton-chops that made its head look broader than it really was. The tip of the pink nose glistened with moisture. Long white whiskers radiated from the cheeks. The eyes were narrowed like two button holes in a fur coat.

The cat remained frozen, but alert for a while, then checked around by slowly turning its head

from side to side while its ears twisted back and forth
to scoop up sounds worthy of further investigation.
Suddenly the head disappeared behind the sticks that
had hidden the rest of its body, but in the next instant
the cat was on top of the wood pile in full view. Again
it cocked its head to see a movement ahead or hear a
sound that could lead it to a meal, but with quickly
acquired indifference it sat down on its haunches and

began trying to rid itself of a thorn. For a little while
the cat sat there working on its paw or looked around
as if not quite sure whether to have a snooze or to
continue to prowl for a while. Finally it got up and
moved along the shore through the grass and the
shrubs as noiselessly as a butterfly. Its movements
were as smoothly polished as a tiger's and it seemed
more to glide than to walk. Its thick grey coat was
streaked with brown and had no markings, except for
the black-edged ears and the short tail that looked like
it had been dipped into black paint. The underside of

166

its belly was a light gray, that changed into white just like the underside of its legs. The feet were heavily padded and seemed to be too long in proportion to the rest of the body. The lynx's large feet enable it to walk on top of snow in the winter so it can hunt the varying hare which is the mainstay of its diet. The hare population increases over a period of ten years, followed by a mysterious abrupt and sharp decline, after which the hare gains in numbers again. This cycle is closely followed by the lynx whose numbers increase and decrease in ratio with the varying hare population.

Other animals occasionally preyed upon by the lynx are grouse and small mammals, such as squirrels.

In the wilderness the message of death is often carried on silent wings or feet. Like all members of the cat family, the lynx is a creature that has perfected soundless travel. Aided by an extremely keen sense of hearing and sight it can pounce on an unsuspecting prey delivering death swiftly with its spike-like extendable claws and its dagger-like canine teeth. Even a small animal like a mouse, that foolishly delights in hearing its footsteps as it scurries through dried grass, is approached with the same skill and care that the lynx employs when hunting larger prey.

I watched the cat suddenly flatten itself to the ground to blend into the surroundings like one of the many rocks that lay scattered all over. For several minutes the animal remained motionless and peered intently through the shag of grass. I trained my field glasses on the area ahead of the lynx, yet nothing caught my attention. But the cat's hearing was reliable. With an immense leap, it suddenly catapulted forward as if forced by a steel spring, and with outstretched paws and extended claws, made a precision landing on its prey. With the mouse pinned down by one of its front paws, the cat seemed almost bored as it looked around nonchalantly. Everything was quiet and devoid of other creatures that were of interest to the lynx. It shook its paw as if it had stepped on some unpleasant substance and the mouse

167

dropped to the ground where it lay lifeless. The lynx batted it several times then scooped it up, flung it into the air and leapt after it. For a while the mouse was tossed and played with, then the cat settled down to a swift meal. A few moments later the thicket swallowed the lynx, leaving me the memory of the wilderness's most mysterious animal.

At home in the mixed forest of the Canadian bush that stretches from coast to coast the lynx has no natural predators, but is heavily trapped by man. It is not as prominently marked as its smaller cousin the bobcat. At the end of April or during May up to four kittens are born in a rock cave or hollow log. The young live a well cared for life for three months; then they are taken out into the wilderness to receive instructions in looking after themselves.

A wild cat that knows how to prowl the wilderness like a shadow, shuns the closeness of man and hunts mainly during the night, easily acquires a reputation that consists of a mixture of truth and fiction. The early explorers believed it to be 'a catlike vampire, that rides through the wilderness on backs of deer'

The Indians explained the lynx's love for secrecy with a legend that deals with the world of long long ago. Then animals and people still spoke the same language and did not have such well defined places on earth as they have now. At that time the lynx had a beautiful long tail, much like the cougar, and was married to a skunk. He was extremely lazy though he had a ferocious appetite which kept his wife busy preparing meal after meal. When all was eaten the lynx would while away the time, curled up by the fire instead of going out to hunt. In time hunger pangs woke him up and pretending to go hunting he left camp. He had no intentions to go on a strenuous hunt but thought of eating the skunk. His wife had grown suspicious of him and hid herself. The lynx searched for her and grew hungrier and hungrier. Finally, unable to withstand the hunger pains any longer, he sat down and began to bite at his long tail. Before long he had bitten off big chunks because he liked the taste of his own flesh. When only a short stub was left, the skunk who had watched all along, could no longer contain herself and roared with laughter about her husband who was so lazy that he ate himself. The lynx suddenly realized what he had done and seized by an immense feeling of shame over his greed and laziness went into hiding. Ever since he hunts at night, because he is still ashamed to expose his stubby tail in fear of being ridiculed.

THE WILDERNESS LOVER

Surrounded by hills that shone in shades of purple and gold in the mild autumn sun the old handmade cabin stood hidden from view among the trees. A few feet to its side a creek gurgled into the cove of a lake that lay without a ripple, content to reflect the sky and trees.

A raven cawed in a tree top and from far above wild geese honked across the evening sky. A falcon shrieked after prey and a blackbird fluted hidden in shrubs by the shore. Somewhere a loon whistled briefly.

This was the world of Everett, a prospector who lived alone in the bush. I discovered his abode as I paddled into the cove for a cool drink and he invited me to share a cup of coffee with him. I knew I had met a man who thoroughly knew the wilderness. His hair was thin and white and his face carved by many experiences. On the back of his hands the veins

seemed to burst through the parchment-like skin, but his arms were strong, still showing firm muscles. His eyes shone with happiness that he had found in a simple way of life, and though he had been a loner for almost forty-five years, he had never been lonely.

He had built this cabin many years ago but it began to succumb to the burden of age and the steady encroachment of the bush. The wind kept prying open cracks and lifted shingles, so that Everett had to place pans and cans on the floor when it rained. The pans were for washing gold, the tin cans used to hold his tobacco.

Panning for gold and smoking were his most persistent habits. He spent an even amount of time on each. He never smoked without thinking, and never made a decision without first having a smoke. He called that 'To smoke that one over'. He agreed with Thackeray's thought of 'smoking draws wisdom from the lips of the philosopher and shuts the mouth of the foolish.'

His cabin was full of gear. There were necessary items: a stove, a table, two stools, a rocking chair, a bed, some shelves. Boxes, books, rock samples, gold pans, cooking utensils, and clothing cluttered every surface. Leg traps dangled from the ceiling and from the walls. In one of them a spider had set up its own trap. Rust, dust and spider webs told me that the traps had not been used for a long time.

Indeed, he had forgotten the last time he brought home a trapped animal. Why did he stop? The question only drew a response from his pipe.

In between puffing on his pipe, he sipped his

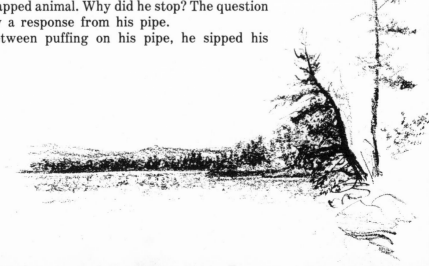

coffee and listened to the silence. So I listened with him.

Flies, spiders and ants all found a home in his cabin. A pack rat rustled under the bed. It was a good year for wasps. They found an easy opening in the screen window that did not close tightly. Nothing fitted any more, yet everything seemed to be the right way.

I questioned him about life in the bush, about trapping and prospecting. Since he was often given to long silent pauses and I was not in a hurry, we spent much of our togetherness lost in thought. When Everett finally opened up, the fire in the stove had gone out, the sun was about to go down behind the hill that nobody had named yet, and the water of the lake shone like mercury.

His traps reminded him of dozens of angry, frustrated, begging eyes—full of agony. They were visions of death.

He had never felt much for the animals: how they were trapped, whether the leg traps meant torture to them or not. He needed their fur and his main concern was that the traps caught plenty of animals. His trap line used to be long and rewarding.

He caught beaver, mink, marten, fox, wolf, wolverine—even bears—and sometimes amimals that

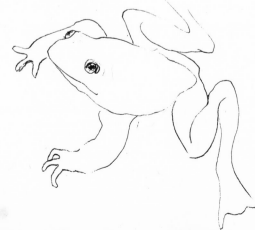

were of no commercial value because they got into the
trap line by mistake. In those years the sight of blood
meant nothing, the knife was part of a skillful hand.
There were no doubts about the animals' place. They
were either food or good for pelts. Life was hard, but
healthy and above all it was good.

Somehow it changed.

Animals disappeared. He had taken too many
beavers and one summer there were none. The
wolverine were gone. Lynx, too! There was the odd
fox, but it was more and more difficult to catch them.
Often the trap lines brought little or nothing.

Once when he checked his trap line he found a
coyote in a snare, and realized he had forgotten his
gun. He watched as the entangled animal stretched
and twitched, trying to escape. It had gone around the
tree so often, it had tightened the wire so it could no
longer walk. Its twisted body lay sideways; the legs
were trying to find the ground to stand on. They
jerked and scratched the air, they dug into it and
pounded it. Life took so long to leave and caused much
agony. He could have clubbed the animal, but he just
stood and watched and watched.

The impact of death by torture and agony was
great. There was an awakening of a feeling he had not
known before. It was so strong that it made him quit
trapping.

To live in the wilderness was his life: this he could
not change. He moved, built his cabin and took up
prospecting. There was much to do, much to learn,
much to read.

Since then he has thought a lot about life and found less need for talking.

He used to hike quite often to the village that was twenty-five miles (40 km) away, but later he went there once a month to replenish his supplies and have a beer with an old friend. Sometimes an Indian friend dropped in. Then they shared some time in quiet companionship or listened to each other's stories. From this friend he heard the legend: 'How the Changer of Things instructed the animals to be useful to people.'

The world was without people for a long time, until the 'Changer of Things' decided that people should walk the earth. He called all the animals to a meeting and told them of his intentions. "These humans," he explained to them, "will always live near lakes or rivers and all of you must decide among yourselves what you would like to give to them." While he was busy creating people, the animals discussed how each of them could be useful to this new creature, called man.

"I will live near him so I can be his food," the hare said.

"I will be his food also," agreed the moose, "and from my hide he can make clothes and shelter. From my bones he can fashion tools."

"I will provide him with blankets," said the bear.

The dog suggested that he could help man by warning him of dangers or going hunting with him. Each animal had something to contribute and pledged his help to man.

'The Changer of Things' was pleased with the animals' willingness to aid man to stay alive and turned to address his new creation: 'Forever you must respect all living creatures for they will be your food, your tools, your shelter. You must never abuse them nor kill their young. You must only take what you need and use all you take. Then your needs will always be met and you will live with happiness.'

When Everett was finished with the legend it had grown so dark I could barely see him. In silence we shook hands and I went to my kayak to paddle back to my camp.

174

Several years later I was again in the area and sought to renew my acquaintance. Armed with a tin of tobacco I walked up to the cabin from the road that had recently been cut through the bush. From the lake came the howl of speedboats, and I heard people yelling and howling in the cottages close by. I had passed a tourist camp earlier on and had continued with mixed feelings. These were no longer the sounds of Everett's wilderness. When at last I found the cabin, I knew he had left. The door had been pushed in and the window panes were broken. The roof had partially caved in and shrubs had climbed over the walls. When I went inside a squirrel fled onto a broken shelf and, clinging to a log that once held up the roof, chattered with excitement. In one corner a chair sagged under the weight of age and a coffee pot lay rusted and dented on the dirty floor. Otherwise the cabin was empty except for bits of rags and papers that were piled on the remains of a mattress. I picked up a page that seemed to belong to a diary, but the thoughts that Everett had put down in his neat handwriting had been washed away by the rain and the pages of his book had been scattered by the wind of a changing wilderness.